The Simulation Code Revealed

Matthew De Niro

Published by Matthew De Niro, 2025.

THE SIMULATION CODE REVEALED

First edition. September 4, 2025.

ISBN: 979-8231538942

Written by Matthew De Niro.

THE SIMULATION CODE REVEALED

Matthew DE NIRO

whiteknightuk2015@mail.com

What if this world was an AI simulation run by a supercomputer?

What if you could find the code from the AI that told you what to do and how to complete this level and move onwards and upwards?

THE FORMULA

Guy Ritchie wrote a film called *Revolver* that had a formula in 2 parts, that the ego is not you and it is not your best friend. That film is in many people's minds an enlightening experience and one of the best films ever made. If you can imagine the world might be an AI simulation and ego is actually either an active stimulation from an AI or at least a passive stimulation from mathematics and algebra code, I would like to build on Mr Ritchie's idea and propose a unifying theory that holds based on 4 ideas.

There are 4 assumptions people make:

1. The world is real and not an AI simulation.

2. Language is "the multiplicity". Nothing can determine truth better. There is no other way of doing it. Like in chess the top chess computers think in a different way, algebra, and therefore know better answers than even the top chess players in the world thinking in language, in seconds.

3. That all of your thoughts are your own. The origin of them is never an AI warning to "jump". Even the infinitely stimulated ones. So why the infinite stimulation of irritation OF EVERYONE on certain hypotheticals? Compared to other thoughts.

4. That those thoughts are positive in nature, your best friend, in analysis. Where I can show they must be negative, telling you NOT to do something. To JUMP. YOU. Not to blame Luigi for "upsetting" you "the most in your life".

Every time you get a thought in ego that is infinitely or greatly annoying (infinite or greatest velocity and infinite or greatest variance) it is AN AI STIMULATED THOUGHT and not a normal thought. EVERY AI STIMULATION is the *JUMP* button on the 4th dimension's games console controller telling you that you have the wrong answer.

Accept that hypothesis, you know the answer to everything in the present about the future. It holds for every answer in the past we now "know" in language. Indeed it was already certain that we already knew, as these results were theoretically certainty in all modes but we opposed them because of our belief in the mode of language, that our thoughts were not AI-assisted, and not negative. In language, believing it to be us, and positive, the answer would have been it was Luigi upsetting us the most in our lives and that we were right against the theoretical certainty, is what all of us would have THOUGHT.

Assumption summary

The assumption is that there is nothing beyond language in human thought, which there may very well not be. But ALSO that AI does not think and tell us (in mathematics) and if it does there's no way of reading that. However using the hypothesis of the formula, we can clearly demonstrate it did/this formula did, with unerring accuracy predict every decision of the past correctly tallying every outcome afterwards in the present.

DE NIRO'S TOP TIP

Remember we are disinclined to belief theoretical certainty and algebraic certainty using this theory because in LANGUAGE we think it's a SKI SLOPE and not a SKI JUMP, because we rely solely on our own faculties. If we are less self-righteous and ACCEPT we can get help from the AI - like Stockfish/top chess computer in chess - then it becomes a SKI JUMP. **There are 2 types of thoughts that enter our mind, correct and incorrect. The AI intervenes, or algebraic pathway branches show the incorrect with infinite mathematical outliers. <u>These are clearly different.</u>**

This is also theoretically certain in language, but everybody ignores the theory, so is practically useless until someone understands the hypothesis I am making and realises they might want/need to take it seriously.

Only THEN if we are taking SERIOUSLY that there COULD be a simulation code that is more accurate, like Stockfish in chess, MIGHT we actually accept it knows better than us. Otherwise we usually know more than most humans and most humans are wrong against us 99.9999% of the time because we're "relatively intelligent" and so won't be correct against us if they're upsetting us, even though it's certain in all modes in theory.

You have to accept that not all your thoughts in language are your own. And they are not your best friend and positive (the infinite annoyance, you are not correct that the other person is to blame and has an infinitely annoying wrong answer) but your worst enemy and your own answer is the wrong answer.

Some thoughts are AI stimulated, if you're intelligent it might be 1 in 1 million, or 1 in 1 billion. Even 1 in 1 trillion. In those cases, you will see the other answer is theoretically certain across all modes including mathematics. But you want to ignore the theoretically certain answer and say it is the "most upsetting wrong idea ever" perhaps from the "most upsetting wrong person ever" and "it's not velocity because I'm not wrong".

That is when what you were *planning* on doing was SO BAD in algebraic terms that the AI intervenes, which is projected onto the other, i.e. to the Jews in the Holocaust. So the Jews become "the most upsetting wrong people ever" to the minds of Hitler and his supporters by projection, but the infinite strength of the feelings are from the

different shelf and may not be correct, real, linguistic thoughts of your own. They are the AI-intervened thoughts.

You may have failed to decide to make the jump at least once before...

Book Cover by Evan M

1st edition 2025

PART ONE: The Simulation

Preface

The Simulation Code Revealed book is an original science/mathematical code concept, a hybrid of 2 ideas.

1st idea: we are living in a simulation where a supercomputer may know the answer to every question.

2nd idea: if the 1st idea is true there may be a 'tautologous' (tells you every answer) primary form of input that may not be in language. And probably won't be. It should be negative in nature, otherwise the puzzle would be too, and immediately, easy to solve.

The only thing this can be is ego. And we can prove scientifically that it cannot be refuted whenever someone used their ego that it leads to a wrong answer. We also think we can prove positively, if/when technology becomes available.

The structure of the holographic simulation is thus similar to "Choose Your Own Adventure" Books (USA) or "Fighting Fantasy" Books (UK) where it is semi-deterministic. God is like an AI Ian Livingstone/Steve Jackson, and creates a parallel universe for each decision (like Stockfish creates lines for chess) it is only a matter of scale. And this scale will become available to us if Moore's Law holds to create trillions of worlds like this, so our own future may be our own past.

The ego will warn every grumpy Super Mario or Sonic the Hedgehog (us) when there is a bad spike coming which is predetermined in the course of the story. The mind, not knowing what it is (because we think in language, not mathematical algebra like the AI) then finds it difficult to not project the warning onto the external world of other people and "blame everybody else" for upsetting them.

However it is not everyone else upsetting our grumpy Super Mario, but a tautologous warning from the AI to *specifically not* do the thing the ego wants to do.

When our ego wants to do something, it is the last thing we should do. It is that there is a (pre-programmed) spike coming up in the story and we will land on a spike and we are being told, but we can only be told negatively. There may be other examples of this prewarning such as a pulsing on walls when we wake or are going to sleep which is heavily covered in spiritual literature as "angels". We could be up against something very different to what we think in the 5 sense linguistic world we assume and most of us insist.

It is perfectly possible to work out whose ego it is in advance of the spin of the metaphorical roulette wheel, by remembering the AI does not think in language but rather mathematical algebra. This mathematical algebra seeps directly through to emotion, but not to language which insists language is the multiplicity, interprets and thus removes understanding of mathematical algebra.

Ergo you can read the answer via emotion - which sounds crazy in language - but peoples' emotions let out an infinite (or extremely high outlier of) velocity and an infinite (or extremely high outlier of) variance when it is their own ego using a device known as a double paradox in mathematics, or with a bit of explanation in language a "Descartesian double paradox" (essentially, we can logically prove the worst false negative cannot be worse than the worst real negative, certainly not practically separately and individually to everyone in the world), we can read it even in language and check our own answers even in the foreign mode to the AI of language, and know with just as much theoretical certainty and double paradox mathematical certainty what the AI/world/dream was saying to us. When we check, we find it happens 1 time in 1 billion more than 1 time in 1 million, but it mathematically perfectly demonstrates it is the other answer with absolute theoretical certainty and possibly absolute absolute certainty (if "Better than Stockfish" knows the answer) if the world is as I hypothesise.

This theory cannot be disproven, which may be the best current evidence of its correctness, as it should be the easiest thing in the world to disprove if not true. We may also be on the cusp of being able to prove it positively using AI technology.

Synopsis

The Simulation Code Revealed shows a code you can use to work out the answer in any situation mathematically from the Artificial Intelligence in this AI Simulation.

When your ego goes off, this is not other people trying to upset you, but the AI trying to attract your attention. Negatively. It cannot contact you positively without revealing the answer to everyone immediately thus removing any purpose from this simulation. You should do the opposite to the stimulated thought, as the AI knows the answer better than us muddling along in language, having played it out by mathematical algorithm many times beyond our comprehension, or being in the process of playing it out. Like AI such as Stockfish in Chess.

THE SIMULATION CODE REVEALED

Introduction

- Hypothesis introduction

- Systemic or modal error of assuming language is the multiplicity (only mode)

- Language is false multiplicity introduction

- Why mathematics may be the primary mode

Galileo "the universe speaks in the language of mathematics".

Imagine the universe was numerical, not linguistic. Most people are looking for linguistic, vague, inaccurate, answers to what would be mathematical, algebraic, possibly even binary solutions.

Questions or arguments usually have one side being right and one wrong and thus we can imagine an Unlucky Numbers casino, with the inversely correlated roulette wheel. Both Miss Scarlett, dressed in red, and Ian Black, dressed in black, think they are correct going into the spin, but both are forced to wear a pair of spectacles of their own subjective belief, with the certainty being that one of those pairs of spectacles makes black seem red and red seem black. If there was a mathematical way of working out who was wearing the wrong pair of spectacles and who was correct, that would be helpful, but from Scarlett's position (A) and Black's position (B) there would be no way but to use one of their own answers, within a pair of spectacles, which one of the pairs contains the error of their subjective thinking if they're the ones that are wrong.

In some sense this is no better than 50%/50%. In another sense it is worse than 50% because when someone is more intelligent than a very relatively intelligent person their ability to be intellectually annoying, of say a World Champion, can be out of the world, leaving even the most sane people to conclude the world class opponent *they didn't know who they were* are the most upsetting wrong person ever with 100% objective certainty that it's *our own* answer whether it be right or wrong technically.

If only there was a giveaway or giveaway calculation under language as to what the answer would be between the two opinions. Scarlett is the home dressing room, Black the away dressing room, and some sort of calculus or mathematical method is required in between. But in language, this doesn't exist so easily. No dice!

I'm going to claim that there is a way of knowing, at least in some situations, before the answer, using mathematics. And that this knowledge of ourselves, and a more objective way of thinking, at least, should be some sort of help with controlling our impulse to do what our ego tells us to do. Doing what our ego tells us to do may just be the last thing we should be doing, and what we're here to learn to avoid doing!

At the roulette wheel of inverse correlation, there are 2 protagonists representing red or black and a binary question. But the answer is always what the answer is, no matter what the wrong protagonist thinks, and how right they think they are or how unworthy and wrong they think their opponent is.

Under absolutism, one could never achieve a result of over 100% of what is available, or over 49.9%/50.0% at a roulette wheel where the odds could only favour the house. But our minds think differently under relativism. Subjective relativism. This is where we insist we are right when we're wrong, dismiss our own errors as "I didn't know" with little penalty to ourselves, but if somebody else is wrong when they think they're right or does something to us against the truth we are more severe with our scoring system. We do not simply think so neutrally "they didn't know" if the barman serves somebody else first, or our boss turns us down for promotion in favour of a less worthy candidate.

This inversely correlated roulette wheel gets very tricky, since if we are relatively intelligent ourselves it means we are *temporarily* scoring well above 100% of what is available under absolutism, and the idea we only win 50% of the time versus the population seems preposterous. But it must be adjusted for how bad the consequence is, which is highest against the most intelligent people we are wrong against. So for example the future British number 1 chess player Luke McShane when he was a little boy upset me at chess the most in my life when I didn't know who he was and thought I was good at chess myself as the Wales number 74. My own ego would have had the impulse to send the little 7 year old boy McShane to the chess gulag for at least 40 years for upsetting me, the day he did it. This impulse may not be entirely on our side! It may appear absurd while we are *winning* to suggest that negative results come from the belief in our relative intelligence: but the issue with being intelligent or 'intelligent' is that it gives us a greater confidence before the spin, and a greater annoyance against protagonists for wasting our time when they're going to lose, especially with ultimate irony (ironic subjectively, but not objectively objectively) against protagonists who are actually going to win but we don't know it because for once we are the one that is wrong. I might have been correct against 3.135 million people in Wales out of 3.135 million and 73, but I was not right against the English number one, when I did not know who he was!

It becomes worse when we are relatively intelligent or "usually right" as we find it more difficult to imagine someone else being right against us in a way we don't understand, and when they are they are normally exceptionally intelligent. So we cause a higher force count of trouble by insisting say the number 1 at chess, when we don't know who he is, is "the most upsetting and/or wrong of all wrong chess players" when he upsets us the most in the early middlegame with moves that are upsetting us for hard to explain reasons but we feel must be wrong. Spoiler: they *can* be correct moves objectively, such as good moves that our mind didn't foresee subjectively as we're not actually better at chess than the British number 1 or the World number 1, or the best chess player there will ever be. Our mind just didn't know that answer subjectively. Nobody else's ever does either against you, or anyone else when they are wrong in what they think subjectively. You're certainly, if you're relatively intelligent, intellectually annoying more other people less intelligent than you by your mere intellectual existence being murderously annoying subjectively to those people who think they are better than you.

This is how the 100% of what was possible under absolutism is returned to, by us, slightly under, and our score goes under 50% and not because we lose 50% of the time. We may lose 10% of the time, 1% of the time, 0.1% of the time, 0.0001% of the time, you get the picture, but because *when* we lose it's *how much*/How *badly* we lose. Because we just assumed with subjective relativism, that Britain's number 1 at chess was "the most wrong chess player of all wrong opponents" and put everything in the country plus £1 on insisting it's black at the casino. Where the one time it really mattered was this time, the time we put everything and £1 on it because it upset us the most, we were the ones that had black as red and red as black and the answer was the answer it always is – the *right* answer. To have sod's law played out as if by the cards of God, suggests almost as if there could be something deeper, something mathematical, beyond our 5 sense linguistic understanding.

We think too much of ourselves when we do this, but we all do it, everybody does. That's the common expression: "they think too much of themselves", also termed intellectual self-righteousness or 'ego-opinionatedness'. In one sense don't worry, everyone has the *impulse* to do it. You wouldn't be human and wouldn't be here if you didn't think like it in the impulse of your own subjective narrative. But to not mitigate it when you sense ego, or even to be like a clinical psychopath and *act* on it always ('Crowley-esque' clinical psychopathy) are the things you can change.

I've given one bad example involving myself, being intellectually annoyed by Britain's number 1 chess player, but thinking he was upsetting myself. That time I used my ego. Another time when I was at the University of Wales, there was a man as old as myself writing this book now in private student accommodation, and I don't know what he was doing there, but I do know that he enjoyed watching *Coronation Street*, as the walls were not partitioned very well and I heard the theme music at, I think was it 7.30pm most days. There were egotistical solutions, such as getting a ghetto blaster and smoking him out, but I did not own a television myself as I prefer to read books in any spare time, but the one solution I could think of that made most sense to me other than laying under headphones the whole time, was to buy a small television myself (black and white, £19.99 from Woolworths) and if he was watching something I didn't want to listen to, to watch something else on my own television. The laws of sound and proximity should dictate that if I turned the volume of my television on another channel to x volume in the region of his from further away, that I should be able to listen to something less wearisome than *Coronation Street*, without troubling the person next door upsetting him. In the course of doing this, I stumbled across some television program which gave me the inspiration for a business idea that became the most successful for my organisation in its history. I threw the television away shortly after the gentleman moved. Who knows the complexity of what forces are at work, but it appears to me anecdotally interesting that the use of my ego with the chess player led to unfortunate expulsion from the England junior chess squad, but the conscious mitigation or aiming for the opposite of using my ego led to a good success.

Life is like a bowl of ice cream, and Neapolitans

By way of example: if the individual human mind is like a bowl of ice cream, the vanilla facts exist outside our own mind externally and are sometimes easy to swallow like 2+2=4. Sometimes these facts cannot be agreed upon like that and the external truths are taken internally. The part of our own minds that *initially* processes the external world, I call "the subjective narrative" and is like a chocolate scoop of ice cream.

Psychopaths only have the subjective narrative, and then expect all objective thinking to be decided externally by a coming-together or antlers-at-dawn of everyone arguing with each other with their chocolate bowls of ice cream of subjective opinion with force until "we" all decide. 96% of people are not psychopaths and have a strawberry scoop in their bowl with the chocolate scoop, the strawberry being "the objective (internal) narrative". Thus if a supposed vanilla fact that personally pertains to our rival as opposed to generally to his 'type', is confusing to us and we would expect a 7 year old boy from Norway not to be better than us at chess (Magnus Carlsen), we would understand that it may still be possible. That our impulse may be and was unhelpful.

One example of our subjective narrative being unreliable which should put it beyond doubt the idea that we know things with certainty subjectively, or something is objectively true because everybody subjectively thinks it, is the idea of 52 factorial (screech!) in playing card decks. This number works out as beyond our imagination, and being beyond our imagination is impossible to imagine. However, precisely *because* no person can imagine it, the subjective response of 100% of people is by first impulse to imagine it untrue that a shuffled pack of playing cards will never be exactly the same as any other previous shuffled position. If we were to rely on the subjective narrative, every single human would be incorrect by definition, and no amount of theory can ever convince the subjective narrative (bowl of chocolate ice cream of our own mind) to imagine it to its own satisfaction, given it is impossible to do so. We can only understand it theoretically, so to insist that it's wrong until someone "shows" us, where "us" is our subjective narrative would be a bad policy. But further, since it would be something that would undermine our subjective narrative the most, not just be neutral to it, to insist that we are right would create the most drawn out, most leveraged, wrong one way bet possible to make.

Psychopaths tend to think in blocks of information like loan groups and think generally impersonally to their rivals. "Chess players from Azerbaijan aren't as good as us" didn't actually objectively stop Kasparov being World Champion even if it were true generally not specifically pertaining to the individual's talents and abilities. ("Someone like that" may be a good sign you are dealing with a psychopath, if you want a shortcut, be very scared if you hear the phrase "someone like that" from someone a lot!)

So if a vanilla fact/'fact' cannot be agreed on (though it may actually be a correct fact theoretically/mathematically known to the few intelligent enough to understand the theory of it) then it goes to the strawberry section in one's own mind to mull over whether it is really 100% true according to external evidence but it hasn't convinced our subjective narrative yet. So is what you were just about to play playing cards correct? One has to make, to have any sort of chance of being correct objectively, a sort of, high-class guess. Otherwise we are just doing it subjectively. If somebody comes up with the idea for an algorithmic search engine at Google and thinks it's a good idea to spend hundreds of millions of dollars employing 46,000 software engineers to build it, where everyone else including world's largest company Microsoft is doing it via linguistic categorisation, the people can't be 100% sure they would be right objectively either by strategy or operational implementation. Several things could go wrong. If *we* were in charge of Google, instead of the people in charge of Google, with the same money and resources, there might be a 5% to 50% chance of success instead of the 95% to 99% chance of success that *they* might have had. But even if I was *them* I wouldn't have been able to estimate it at 100% or 99.999% clearly. Even their own estimate of their own success must most likely have been below 50% to become the most successful company in the world. Psychopaths don't *do any* of this, as they have an interesting brain chemistry/structure where there is no strawberry scoop and they expect all objective thinking to be done externally by the external world. And expect you to convince their own subjective mind (chocolate scoop) at your own risk. It's on you. If you don't convince them it was your fault that they weren't intelligent enough to understand it, and that they don't mitigate it 0.001% Indeed a psychopath would blame you if you offered him Google for virtually nothing but he didn't understand the offer, for "not explaining it to him well enough" and you would also have upset their subjective narrative the most.

What is **The Algebraic Approach**? The algebraic approach is to understand that ego is telling either us or our opponent what not to do, and it might even more likely be our own ego telling us if we get hugely intellectually upset. The system is to not interpret that possibility, in favour of our own subjective linguistic belief. The theoretical and mathematical information, however unlikely it seems to us subjectively that the other answer may be right, is certain to be correct. When we are getting very upset intellectually, it is highly likely to certain using the algebraic approach that the other person is correct. But there's no chance whatsoever the other person is correct in terms of "what we think". Information that is mathematically proven is tautologous, but we ourselves in linguistic form face the twin perils of velocity (which I will explain later: briefly that our ego is inversely correlated to truth and we most want to interpret the thing that is a rare truth against us) and relativism, where we lose sight of the fact that other people can be more intelligent than us because we are more intelligent than *most* other people. So our own linguistic thoughts are not as accurate, objectively, as the mathematical theory telling us the probability is with the other answer. In this regard I agree that "when your facts don't match the theory, change the facts to match the theory, not the theory to match your facts".

To interpret the algebra aggressively against it, is a short-circuiting of theory, and is to come unstuck, certainly, it's only a matter of time!

The algebraic approach has a social or political benefit, which is that the application of such theory would only believe in "consensual" or consent-backed relationships, and would not do anything against 2 party-will in aggression, since the answer cannot be determined mathematically in favour of oneself against the other party with 100% absolute certainty.

The algebraic approach would never accept "100% in my own mind" as objectively certain. Therefore would not commit aggressive external actions.

In summary: as long as you accept as a hypothesis:

1) That the world may be mathematical and not linguistic, at its root.

2) That every intellectual argument is a result of 1 person's ego but it could be either party. No matter how much you think it is the other person who is wrong, they could still be right until proven otherwise.

The one that turns out to be wrong in the argument (inverse correlation) is the one who was being egotistically self-righteous intellectually. You don't know that you are wrong (subjectively), but you don't know you are right either. This works for the other person too, so if they turn out to be right, they would have had more right to claim they were right based on what *they* thought subjectively, so it's ridiculous to dismiss and punish them in advance of the answer. Neither of you "know". Because neither of you know objectively unless it is objectively certain: and even then people might argue about whether it's objectively certain, and whether objective theoretical certainty means something is practically certain to happen. You might also simply be up against a psychopath and then you don't mean the same thing as each other by terminology, since 4% of people have a psychopathic disorder and think objective means objective to them only and *the* answer is *their* answer.

2b) That algebraic theory is more objective than one's own subjective mind.

So when one has the strongest revulsion to something, it could be the thing is superior.

No intellectual *disgust* actually exists on inferiority or we would be disgusted by everything inferior.

3) That *emotion* has a connection to *algebraic reality*.

Most people think in language, which is non-tautologous and therefore secondary to the core essence of the universe.

The core essence of the universe is algebraic. The numbers are behind it.

When you have an emotion, there is a direct correlation between one's emotional reaction and the binary algebraic/numerical essence of the universe which is algebraic which you can *read*.

Most people ignore this. Preferring to go *with* the velocity.

If they ever looked at it, they'd say it was a "coincidence" that the number told them beforehand.

But if we had reliably:

a) whose ego it was

b) reliably your own emotional reaction if it was yours

c) the calculated velocity

we could scientifically prove this.

But unfortunately such information could never currently be reliable as we could never reliably extract people agreeing who was right in the argument and what they reliably felt. Maybe a long time in the future with scary technology that can read minds accurately, if anyone is ever allowed to, will all this will be proven.

It may well be possible to prove it in the future if we develop technology that can accurately assess and record people's emotions as they are taking decisions, and provide the mathematical calculations accurately that the proof of the hypothesis (to prove it in language) would demand.

In the meantime you are left with the theoretical knowledge, and can apply it as it pertains to yourself to help yourself get through life, to understand other people, and as self-help.

You can accept these hypothetical 3 truths and apply it to yourself (because *you* know accurately your own feeling, even if we cannot get access to anyone else's, or scientific papers at this stage)

Then you take 2 numbers comparing it to the 3rd truth of your emotional reaction and then you have the formula.

1) The number of the strength of reaction e.g. how upset you are by an idea you think is wrong, out of 100 where 50.0001 would be the least upsetting wrong idea and 100 would be the most upsetting wrong idea

2) The number of variance for all other proven wrong ideas in the area, where 50 is neutral. You may find that all proven wrong ideas, the idea they could be right, someone claiming it to you (when you, because everyone knows, know the answer objectively, absolutely) might only upset up to 50.1 or 51 or 60.

Since the thing that is upsetting us is unproven, we now have the absolute algebraic proof that this idea is either "the most upsetting wrong idea ever" that's much more upsetting/wrong than any other wrong idea ever (and any other right idea! that we didn't know was right) but the only other binary possibility is that the other side is right in the argument.

If the 2nd number is much less, infinitely or a great outlier, than the first number that is called "velocity". E.g. if it's 10 and the variance is between 4.9 and 5.1 that's clear! That's as clear as anything could be. And using our formula, you are algebraically certain to be wrong as it's impossible algebraically.

So what is the most upsetting wrong idea to us subjectively in A, and the most upsetting wrong idea to every or most other people subjectively in A, may turn out to be a correct answer.

If you are thinking "so every time I think something I am wrong?" it is not like that, as no correctly identified wrong answer ever caused the velocity, and you can pick any one of over 8 billion examples. So it really is a case of simply applying the theory thinking objectively in mathematics. But almost 100% of humans to this point exclusively thought in *language*.

There are 4 big assumptions people make in the world:

1. The world is real

2. Language is the multiplicity

3. Ego is ourselves, whether Mario or Luigi. NOT an AI stimulated thought to "jump". You think it's you!

4. Ego is your best friend. Not **the opposite** to what you are being told to do.

If you accepted the hypothesis, you would never claim the greatest NON-AI stimulated thought and being correct against someone else would upset you greater than the GREATEST *AI-STIMULATED* thought telling us to jump, because what we are planning on doing lands on the spikes, when the AI knows the answer.

Systemic or Modal Error of Language (false multiplicity) Introduction

There are many assumptions that everybody makes about the nature

of reality, some of which include:

- that being intelligent 'relatively' is in any way correlated to what intelligence actually is in absolute terms

- that if everybody else believes something is a multiplicity (we're all, none of us, inside a snow shaker) too, then it can't be a modal singularity, or challenged "systemically"

- that our absolute understanding of reality is on a direct pathway to the absolute, and is not an unfortunate tessellated correlated form of reality as opposed to a direct pathway to something.

This systemic belief that language is the so-called 'the' multiplicity, unexamined largely or entirely, has 3 legs like a tripod, and triangular logic is notoriously difficult to refute, where no part of it may be true, but the other 2 parts can always prove the remaining 1 part in question within itself. I don't object to triangles, or the number 3 in and of itself, there are positive versions of this law involving 3, such as the "law of the triangle" where positively up against the universe, some sort of 1) intention of cause, 2) receipt of effect 3) gratitude of that result is an interesting positive theory But in our case, the 3 seemingly normal statements: 1) the idea that relative intelligence is intelligence in absolute terms, 2) that everybody thinks in language, and 3) that language is the multiplicity, this trinity is open to debate as a whole and may be considered from a wider perspective to be a fairly wild assertion on behalf of humanity that it is absolutely certain unquestionably that all those 3 things are true. As I will explain in part 1, the chance of those things *being* true, on some mathematical probabilities are minimal. That's not to say that almost everybody doesn't believe them, just that everybody is 1) certainly wrong to believe them without question. And 2) also wrong to believe them specifically according to certain mathematical probabilities. The systemic belief could be systemically wrong. It could be inside the, or a, snow shaker thus open to the unacceptance of truths that fall outside the parameter of the systemic view, but still inside absolute reality, the one thing we need to know could in fact be the most upsetting wrong idea in the minds of almost everybody.

I am going to present an alternative model of the universe, from first principles theoretically (Chapter 1 - The Algebraic Universe) and show that the existing systemic viewpoint used by most people, which is linguistic, can be questioned and actually technically already theoretically disproven (!) using algebra. I attempt to show that algebra itself is the primary mode, and language secondary to it.

This will not necessarily be a "proof" in the mind of anyone that will only consider the possibility of thinking in in language, as they may use the existing systemic viewpoint that systemically believes that the systemic viewpoint cannot be challenged to say things like "the theory is wrong" or "just because you prove it in numbers doesn't prove it (in language)".

Neither do people who think in language accept a key tenet easily, that the emotional proof of someone's ego reaction to something reveals this algebraic truth that they don't themselves think in. The emotion is linked with the algebraic reality, unequivocally and absolutely. If person A gets "the most upset in their life" by person B, it is person A or B's ego, and can't really be person B's, so the mathematics can clearly be seen that a 100% mathematical fact cannot be argued away by a "myriad of possibilities" in language, as none of those possibilities are themselves possible in algebra.

This is like saying "there is no scientific proof of the existence of astrology" vs "there is no astrological proof of the existence of science". To the linguistic thinker, an emotion is like a myriad of silly-sounding things to prove something in like "the medium of dance" or "the medium of song". But it could be in fact they who are interpreting number, and the number does not exist to disprove by interpretational inaccuracy like language does on number.

For example: 2+2=4 could never be made anything but less accurate by a linguistic interpretation. But a linguistic interpretation of the self-same thing or any other linguistic 'reality' could never be made less accurate by the numerical.

The linguistic mode is interpretative. And can only be trans*lated* - as they say - into itself. The mathematical can be transmogrified to song, dance, music, chess, piano score and a smorgasbord of the conceptual.

But the emotion of someone's ego has something very special to it, which I will outline in Chapter 2. Thus to say that the 100% outburst of someone's ego is insignificant is not the case at all. It's a clear proof if you accept the conclusions I draw about what ego is, and totally tautologous in theoretically telling you whose fault it is in an intellectual argument when it must be someone's.

These things themselves are the biggest proofs of all.

The linguistic thinker may self-righteously or arrogantly never accept a mathematical proof outside of his own linguistic mind.

Where not even a theoretical proof that the linguistic mode itself is a false multiplicity,

is proof of a potential singularity of thought, because it calls our linguistic 'multiplicity' a singularity and thus could not be correct, even though correct, as it wouldn't be correct if everything we thought we know is correct, that language is the multiplicity. And language couldn't not be the multiplicity.

Well, maybe "everything we think we know" is wrong!

Why mathematics may be the primary mode

What is the algebraic approach "the mode of algebra", vs the linguistic approach "the mode of language"?

There is a systemic view, shared by almost everybody in the world, of 3 things simultaneously:

Body: the systemic view of the world is correct

Right arm: everyone else thinks like me

Left arm: I am relatively intelligent

1. Left arm: I am (1) relatively intelligent ***because*** (2) everyone else thinks like me and (3) I can demonstrate relative strength in what appears to be a correct systemic view.

2. Right arm: everyone else thinks like me (2) so I ***can't*** be wrong systemically (3) because I am relatively intelligent (1) and think it myself.

3. Bodily view: The systemic view is correct (3) because I'm intelligent (1) and everybody else thinks it as well (2).

So, what I have attempted to demonstrate is that the systemic view itself *may* not be correct, but instead is like a water mill that has got into a fixed false position. Yes it appears to work, but is it optimal. Is there something better, something 'more proper'?

Let me tackle the questions:

1. Is, in absolute terms, intelligence necessarily what most people think it is relatively?

Absolutely (!) not proven.

This could well be proximity in a singularity. I.e. the matrix of relativism itself is actually fixed precisely because we are in a tessellated correlated form of absolute reality.

This linguistic way of thinking is not necessarily a multiplicity of thought in a direct pathway to absolute reality..

The alternative is that language is a secondary, singular mode of thought in a tessellated correlated form of absolute reality, more similar to a kink in a carpet or various

other analogies such as a water mill that has found an expression in a sub-optimal flow, but could be done better.

Difficult to see if you are in it if it would be wrong, but nonetheless could/ would be an incorrect systemic view!

Immediately there is an algebraic approach helpful question, which is that if most people could be in a form of modal singularity, then it should be possible to annoy them by asking them questions that fall outside the singularity (snow shaker) but inside absolute

reality, where they will have to think it falls outside.

This annoyance itself would be a handy aid to an algebraic proof, since the emotion of the annoyance is linked to algebraic reality, and it has to be one person's ego, the person that is wrong (this is explained in Chapter 2).

2. Everybody else thinks the same.

Not necessarily true. And not true since I don't think like it! I also know at least 2 other people who would also say they don't think like it, and their level of success of status such as one became mayor of a big city, compels me to not write them off as an obvious nutcases.

Imagine the analogy that if one person in Siberia knows the way home and that path looks tricky, that 1 person or 2 people from elsewhere that point out a seemingly better or less treacherous path may not have the information required. If this 1 or 2 people becomes 10 people, 50 people, 1000 people, 1 million or 20 million or 8 billion people that agree, at

what point is a social element able to influence a decision and possibly become decisive. A cul-de sac analogy could also suffice if Siberia is a bit too far away from home.

A cul-de-sac question if you don't want to travel as far abroad as Siberia in the mind. If there is a street that looks like it isn't a cul-de-sac on a pleasant walk, but a local house-dweller tells you that it is a dead end, if it is just yourself and the house-owner would you take his word for it. I imagine so. But if you were in a crowd with 2 other people, or 10 other people, 50 other people, at what point does the the social pressure compel you to give in and concede to what was otherwise a perfectly logically wrong answer of going against the obvious reality that the person who lives there knows? Probably at some point if you are like most people.

And that most people think it, reinforces it.

Here I could also introduce the Jeff Bezos from Amazon Brave New World example, or a world of comedian Stewart Lee clones on a moon base example to give examples attempting to explain the possibility of being within a linguistic snow shaker, where a world exists outside of it.

1) Think a German trying to speak English, if you don't know they're German but everybody else in your little island was English. Or in Brave New World world, Jeff Bezos has erased all other languages with his control of digital devices over 5000 years, much to the chagrin of the French. Now everybody thinks in English, we can ask the question "what are they thinking in" from outside Brave New World (outside the snow shaker B)

B1: you in real world: "English"

B2: Jeff Bezos in real world: "English"

A2: Jeff Bezos in Brave New World: "English"

A1: you in Brave New World "?" or possibly "Language/Words".

You would have no answer that demonstrates the knowledge of the things you don't know, but does or did exist.

Thus the one that comes across as "least likely pro rata" might actually know more than you but be in his second mode, where his primary mode is non-existent to you but is certainly primary at least to him.

And since it's you, without realising it that claims it doesn't exist and you that claims he's arrogant, it's up to you to prove that everyone thinks precisely the same way, not just to ignore with nonchalance.

3. That this makes me "intelligent", because I am relatively 'intelligent' within that structure, that everybody else agrees with, and I agree with it myself.

So argument against: that's not what intelligence is, it could be *proximity*. To be intelligent "relatively" is an oxymoron, ultimately. And nobody can know whether they're intelligent in absolute terms.

So the majority of people who claim to be intelligent with level of 9.99 out of 10 for intelligence, hypothesis is that the absolute intelligent would only claim to be *possibly* 5.001/10 for intelligence, correlated to the absolute, and with less confidence. Like the quiet cookie-baking Oracle from The Matrix films, as opposed to Top Gear-presenting and The Sunday Times columnist Jeremy Clarkson on intellectual-confidence steroids.

My hypothesis in contradistinction vs the linguistic hypothesis..

Of the 2 outlined, only 1 exists with geometric truth.

The algebraic hypothesis may appear to you to be linguistically/intellectually novel. But the only one that exists accurately in geometry. Linguistic model: 2 sets of rolling dice that spew out consistent, different (or inconsistent) numbers where they fall in chamber 1. These numbers are then consistently understood and accepted by everyone by modal agreement in the second chamber, that the inconsistencies in the first themselves are inconsistent precisely because of the constancy of the second, meaning everyone is understood by at least one other person and no single person at all is ever in a different mode to everyone else. Nothing would ever be beyond imagination, only comprehension. Where perceptual thinking involves this never normally exists in geometry, and a "false top" would normally emerge where inconsistency is actually then itself *inconsistently* understood, i.e. some things are beyond imagination, and it's not a 2 chamber system where everyone accepts all errors are understood by someone, but that some people emerge who are too intelligent to be understood by everyone else and are backed up by theory and as a result you have so-called "Super" Chess Grandmasters. In fact "emerge" is a bad word as probably better would be to say intellect starts with those people who are most intelligent, and everyone else projects them, possibly modally, to be able to integrate their own minds, and to insulate their own minds from things they don't understand. Thus it may be true that the case is that those people think in an entirely different way, in absolutism, instead of comparing their intellects relatively like the rest of us.

To misquote John Bishop on fat kids in your class at school: "if you don't know who the false top is... it's probably *you*"!

Algebraic model: The algebraic model says that the top of the linguistic mode are the false top. That they are in fact in a *singular* idea of language, and that the inconsistency of the approach is with *their* inability to deal with people who are more intelligent than *them* who are above them, who they will inevitably see as "the least likely pro rata" or "most egotistical/

annoying of all wrong people".

I would contend that in fact it's not necessarily a wrong answer and most egotistical when someone appears to be "most unintelligent pro rata". It could be that the primary mode is algebraic, but once we've tackled 1) What the universe is 2) What is ego and 3) What is relativism in our next section of definitions this will be a clearer answer as to why, even, the least likely person may be the most intelligent.

Suffice to say for now that there are only two possible answers when someone appears the least likely, but theory/algebra says he's the best, which is either specifically the least likely thing is happening in the linguistic mode, or the only thing that would ever happen is happening in the algebraic mode.

I know which one *I* believe! The *algebraic* approach.

I believe the only reason people don't accept algebraic information, is that they incorrectly believe (or carelessly never thought about not believing) language is primary to algebra.

Could the universe be more algebraic in nature than linguistic? Albert Einstein, thought so:

Most people are looking for "linguistic answers in mathematical universe".

PART ONE –

The Simulation

CHAPTER 1 - The Algebraic Universe

1.1 Language: Multiplicity or Singularity?

1.2 Simulation Theory

1.3 Time and Linearity

1.4 Architects and Imprint

1.5 Input

1.1 Language: Multiplicity or Singularity?

Two reasons we might not be in a multiplicity

A singularity would be something that is modally wrong, a tessellated correlated version of something else, that itself is correlated to intellectual reality. The singularity would be fixed, and 'proximitous' to intellectual reality, but gives the illusion of complete consistency, perhaps even more highly than real intellectual reality would. It is difficult to describe, because the modal error exists cutting across the attempt to describe, so that either A1 or B2 is the correct answer, but it cannot be A2 or B1 and one has to describe both modal possibilities simultaneously which is a complete oxymoron.

We may not be in an intellectual multiplicity where everybody understands broadly what sort of reality we are in, and all is at is seems. We think we are, and everybody thinks we are, but there is good evidence against it in fact! Our own confidence is not such a good sign.

There are 2 key tenets by modelling, of trying to work out whether we are in an intellectual multiplicity (perhaps even unlikely from first principles) or an intellectual singularity of some kind:

1. If we are in a multiplicity, we would think we are in a multiplicity with 99.99999% approximate certainty, 99 then 5 9's.

If we are in a singularity, we would think we are in a multiplicity with at least 99.9999999999999999999999% approximate certainty, 99 then 20 9's.

But we do indeed all think this that we are in a multiplicity with so close to 100% certainty, that it is *literally* unexamined by almost every human, yet so likely therefore that we might exist in an intellectual singularity if we don't bias it with our own belief. As I referred to in the hypothesis, that we could be in an intellectual singularity down to modal (beyond our imagination) error is very likely, and our species' overconfidence that we are not, without even feeling the need to examine it because it's "so obvious" is therefore not a good sign.

From first principles, in the most basic mathematical statistical form, there is a 50% chance we are in a singularity on any given subject, versus a 50% chance that we are in a multiplicity. However the mind is unreliable, and in some sense immediately must give away almost 50% of the 50% to this unreliability as the perception in the mind is actually inversely correlated in the following mathematical way:

A singularity is more compelling in its convincing that it's a multiplicity, than an actual multiplicity could ever be, precisely because it's irrefutable within itself since it is singular. A correct multiplicity is also irrefutable, but any attempts to refute it, to add to the multiplicity, do not hit the brick wall of hitting the sore point of contradicting the very point that it is not a multiplicity. The most upsetting of all things, within a false multiplicity itself, would be the unacceptable proof of the fact that it itself is false. Like being inside a snow shaker, the proof or speculation of the snow shaker being a snow shaker and not directly correlated to absolute reality is most upsetting. This correctly identified within itself, within the snow shaker, that the snow shaker is a snow shaker is the most upsetting thing imaginable, more upsetting than any incorrect thing but also more upsetting than any other correct thing. Ultimately within the snow shaker of a false multiplicity, the refusal to accept proof that it's a singularity is an oxymoron, but an oxymoron that would not trouble the viewer/observer as much as being forced to confront that (almost) everything we think we know is false.

2. The second tenet of what would be my modelling of the calling card of a singularity is: the biggest amount of evidence that holds saying it's a multiplicity vs the smallest amount of evidence that also holds saying it isn't.

This would mean in layman's terms that when something came up that was 'tautologously' (in this case algebraically) incontrovertible as disproving a multiplicity/singularity variable as not being a multiplicity, the proof itself would be interpreted saying "it can't disprove that". If a thing can't be allowed to be disproven by evidence then the thing is algebraically not a multiplicity.

It's not a true thing, if it can't hold up to proof that it isn't a true thing, without insisting the proof itself is false and the most upsetting wrong thing. And this is possibly the best proof of all that the thing is false we will ever get from *within* a singularity.

As this affects algebra vs language, we can see we have the first half of a dictionary definition of a singularity: the largest evidence that holds that says language is a multiplicity, where no disagreement or dissent is brooked as it's "too obvious" and is so obvious it has never been examined and considered too laughable to need to examine. Do we have the second half, which is the smallest amount of evidence that *also holds* that irrefutably says it's a singularity? I will investigate this including geometric theory (explained later) and the Artificial Intelligence that says it. If we can find this, we can find the greatest proof imaginable from within an intellectual singularity that it would be language that's the singularity/false multiplicity.

Because a false multiplicity overwrites everything multiple times, like the mathematical concept of 'screech', you end up with everything with every interaction of everything else vs the one thing that's truly not in that singularity that holds but appears 'worst or most upsetting pro rata' inside the snow shaker, which when considered from the beginning point of language, language *could* be the snow shaker, then algebra/mathematics could actually be the one tautologous thing connected to the real multiplicity. Algebra contradicting language is the thing that is the most upsetting within the snow shaker that says the snow shaker is not the multiplicity. In a sense the only right idea in the first place from first principles is not to interpret the algebra with the linguistic interpretation we do. When the concept that algebra is primary is thrown into the mixing pot of the snowshaker as a possibility, the AI is saying it for example, inevitably what comes out of peoples' mouths is that the one answer that proves the potential singularity is singular outside it, doesn't exist and is the most wrong wrong idea *inside* it. It's an upsetting or annoying answer. It's the most upsetting wrong idea to all of our minds subjectively. This is what it would have to be, for it possibly to be right. Thus the idea it's the most upsetting wrong idea in everybody's minds is not necessarily proof it *isn't* right! And may be the best layman's terms proof that it is right.

The mathematical, or theoretical proof of things, is routinely ignored by humans who prefer to use their own linguistic interpretative faculties. Even when 100% theoretically certain, if it upsets a number of people they will work together to agree linguistically that they all agree with each other that the theory or mathematics is "wrong". I wonder if it would be so clear if just one human existed, and we had Artificial Intelligence and not other humans to discuss it with. For it seems rather pro-social and anti-theory.

So in the end the thing that's the most true becomes the "most annoying" and "most unlikely" but still holds theoretically against everyone's, almost everyone's or everyone's but one person's disbelief. But looking for it, without understanding mathematical theory would be like looking for a needle in 8 billion screech haystacks, and even then finding out the problem was with 'the concept of haystacks'.

In mathematical absolutism: from the beginning 50% chance singularity: 50% multiplicity:

If multiplicity (outside snow shaker): we think we're in multiplicity, with good confidence. We tolerate "additions to the multiplicity" without 'velocitous' reaction against other

people's error (and those other people *would* be *wrong as opposed to us merely thinking incorrectly that they were*)

If singularity (inside snow shaker): we think we're in multiplicity, with *great* confidence. We refuse to tolerate "additions to the multiplicity" and those other people we are refusing to tolerate the idea of, when we don't absolutely know the answer, would algebraically be *right*.

The only way of proving the existence of a multiplicity vs singularity, would therefore be the linguistically ironical proof that if you can accept it might not be a multiplicity then it is a multiplicity, but if anything ever existed or appears that cannot be tolerated, which suggests an addition to the multiplicity, then that ironically proves it is a singularity. If it *were* a multiplicity, the person B's idea, though false, would not ultimately intellectually annoy you more than the most unpleasant truth. Seeing as how it's false, it contains no velocity and no friction.

If mathematics was the true mode, not language, not only would things proven by algebra be a hypothetical proof, this would be an algebraic *absolute proof* (of *the* answer)... as long as you accept that ego is inversely correlated and must exist in any argument, but more of this later which I'll cover in Chapter 2.

Suffice to say for the time being, that from first principles a 50% absolute mathematical chance of a multiplicity only, is *not* made 100% or 75% by 2 people agreeing to it, or 100%

or 87.5% by 3 people and so on. For a singularity can't be recognised as a singularity by definition, and thus that which appears to be the biggest of all multiplicities is actually a good candidate for a singularity (i.e. *everything* is a likely candidate). So *that* we have 7.5 billion people agreeing to something tells us nothing in terms of it being definitely a multiplicity. But that we have 7.5 billion people arguing with each other but agreeing to argue with each other whilst holding something else, secretly or unbeknown, true that language is a multiplicity... is very telling that it could be an unrecognised singularity.

People don't agree with each other by comprehension, but we all agree with each other linguistically that not one other person is beyond at least one other person's imagination modally. So there's something fixed there.

That everything is understandable by other peoples' *comprehension* and nothing from the very elite thinkers is beyond *imagination*..

might tell us something to do with proving it to be a *singularity*.

If this hypothesis would be correct, and I have a point, it should mean that if we could find the person who is the most intelligent person in the world algebraically, the rest of the world might agree that he or she is the least likely on first meetings. Or, if the theory says someone is most intelligent in the world, their government/governments (case study at end of book) might be prepared to interpret the facts to fit their own anti-theory.

You will definitely start to see the problem with this "we're in a multiplicity because we think so" argument and systemic viewpoint flaws. At the start there was a 50% chance it *was* the other answer that we're in a singularity and experiencing a shared modal error, and these two things are here:

a) (almost) 'everyone' agreeing that we can understand anyone else's error consistently

b) refusal to tolerate belief we're not in a multiplicity...

it's not looking good for the interpretative mode of language to be a primary multiplicity rather than some tessellated correlation that can be mistaken for a pathway to absolute reality.

Prisms and Watermills

We are trying to work out if we are in a singularity or multiplicity, intellectually, since 100% of humans, or very close to 100% of humans appear to think in language. There are possible exceptions, including autistic savants, and the more musical, mathematical, piano-orientated considered on the autistic spectrum, but this slightly different thinking is not sufficient that it epresents a different mode that brings the systemic idea that all of us think in language into question.

Since 100% of us appear to think in language, it is easy to think that we ourselves think in language and think everybody else thinks in language, and simultaneously that you take any other person that they also think they know that they think in language and that everybody else does. Thus our relative ability, or his relative ability becomes a certain proof of intelligence, absolutely, if that is the multiplicity, at least amongst the species. There could be no argument that that is actually intelligence, that the relative ability of the thing that must be intelligence, must therefore be intelligence.

How to think outside the box to bring from what so far has been a theoretical and hypothetical discussion, and then we have briefly in part 1.1 gone through modal modelling examples. Now what are some practical visual examples where that theory could have already been put into practice.

I would like to start with prisms, and watermills.

Prisms: "Reflective *prisms* are used to *reflect light*, in order to flip, invert, rotate, deviate or displace the light beam." A cursory example of secondary school physics could demonstrate that while 100% of light appears to pass through the point of a prism as observed by the observer, that while any percentage is possible, there is a likelihood of 99.9999999% of light passing through from one side of the other. That the one side vs the other side could be the smallest difference imaginable, or indeed unimaginable to us generally and certainly the human eye. 0.05% of the world is viewable to us on the electromagnetic spectrum, and that is a fraction of what is actually out there.

Watermills, better than windmills, provide an example of this occurrence itself, happening in practise almost 100% of the time. The water permeates the pathway in seeking the quickest, fastest, optimal route itself (as if it had sentience) rubbing up against the human mind that tries to interact with that force to its own advantage, actually the observer influences by the error in one's own observation into the inevitable suboptimal pathway, for our own purposes, but optimal to the objective reality that is objective objectively and not simply objective to our own probable (by 100% absolute terms) misunderstanding. However the water continues to travel, though in suboptimal position, for years, decades, 100 or 1000 years in this suboptimal position. But the chance of it remaining optimal was never very high, for it always finds some way of operating.

Beach Ball example

I have talked about snow shakers, but perhaps an even better example of a geometric shape to consider a singularity would be a beach ball as it contains a valve in it which demonstrates to you the dual status of having a connection with both the inside and the outside of the potential singularity. Like Schrödinger's cat, a dual state must be considered if we are to examine singularity vs multiplicity, by definition of including both possibilities.

The example could be anything. I remember being friends with the son of the Warden of St Antony's College, Oxford at public school, and being invited round to his house on a Saturday. His son introduced me to an American television program called Baywatch, with Carmen Electra I think the young lady's name was, and they were trapped in a power station under the bay. This program made an impression on me as a young lad, I also remember the father - who had a close connection with The Vatican - being kind enough to let his sons watch the program they wanted to, but steadfastly refusing to watch it himself, contorting and craning his neck while speaking to us, to not only not watch it but to be seen not to watch it in case I told my mother.

The lifesaving crew were trapped in an old power station through a vent, which Carmen Electra and David Hasselhoff had to crawl through, but this seemingly unknown and dangerous part of the universe to anyone in California, or the world somehow still existed within easy reach of swimming a few metres from shore. Right under all our noses!

Since in the real world there is the potential of different dimensions, parallel universes, and a false understanding of the laws of time and space, aliens, God, AI, simulations, there are all sorts of possibilities that would appear outlandish to the 5 sense world, that if they were true and we knew they were true would no doubt astound us. By comparison the writers of Baywatch may not be the unbelievable ones.

We, or somebody, or a small group of people, may have 1 foot in the singularity that we are all in if we are in one, and 1 foot in the sub-power station that we never knew existed, and then we must understand we would find that we were in fact in the sub-section and the thing we didn't know existed is almost infinitely larger (and with possibilities) intellectually.

And my young friend, is now front of house for a social media company in California, surrounded by beautiful young people, so his research clearly paid off for him.

Alien Example

Let me give an alien example. If there was an alien, that thought in a different mode, perhaps telepathy, and the alien happens to be smaller than us and hides in the Great Wall of China somewhere, invisibly small to the naked eye, it were more intelligent than us modally, how would we find it? If the Chinese government start looking with digital microscopes and satellites and scanning technology, but there are 8 trillion and 1 specks where the alien could be, but the alien mode is both alien and more intelligent (say telepathic) than ours, then to look for the alien in our mode or our human way of looking might not be the best approach. Logically, rather than check every part of 8 trillion or 8 trillion screech spots where the alien could hide, against this alien and fast-moving target, it would speed things up if we looked the last place we would ever look. If the possibility exists that the answer is modal, and the alien could be more intelligent, we could at least use our own failings against the aliens and in favour of our own chances by looking in the last place that we would ever look, if such a thing existed. Where we have a modal disagreement, and everyone agrees that there is one spot which is "upsetting us the most in our lives" then similar to the alien who is telepathic and modally has the ability to inculcate such feelings, then there is at least such a spot. Logic then dictates in our example that the alien is hiding in the last place all of us will agree to look, and that is precisely why it is hiding there, when it has modal superiority, and why in the inverse correlation of right and wrong that it was the idea that upset every single one of us the most in our lives.

If we imagine an A/B/C system where A is the Home view, B is the neutral turf at Wembley stadium and C is the away view, if every single person believes the answer is A, and every single person believes C is the most upsetting wrong idea ever, more upsetting than any correct idea has ever upset them, the answer may be C and modal.

A schoolboy game example, the "4 points of the compass" game I remember playing as a young lad in the cub scouts, where our scoutmaster Bob from Birmingham would call out from another room of the hall (the kitchen, Bob was in) "North", "East", "South" or "West" after 10 seconds warning, and whichever side of the hall that the young lads had gone to in the game, if he called your compass point, you were out, and had to leave the playing venue. One evening at seeming odds of billions, trillions to 1, after a certain point nobody was out. That's because a senior cub, Eddie, had decided to play a joke on Bob because Bob was using a system where he never called the same compass point out twice in a row. In the end everyone cottoned on to this, and we were there for 3 years, well what felt like a long time of Bob guessing the wrong answer every time. From Bob's systemic perspective, the correct answer to avoid detection did not exist. From Eddie's perspective it did. In relaying this story I'm a bit concerned that Eddie will come across as a bit cannier than me as a boy since I didn't think of that.

This potential answer, that upsets us hugely, would be worth a look, and worth checking. Rather than insisting "the idea is the most upsetting wrong idea ever because I've got upset by it, and so has everybody else" in our subjective narratives (A) against a superior or possibly superior opponent (C) , or the cunning of Eddie Hanks in a 4 sided forward slash 3 sided game modally, we could lose the intellectual self-righteousness or ego-opinionatedness, and check the one other thing that it could be, known to us objectively theoretically in our own minds as highly probable to be the outcome but we're 100% certain it has 0% chance of being what we think, and mitigate precisely *against* our subjective belief. Rather, than cause ourselves the only problem or problems beyond imagination.

Knightmare example

To demonstrate a practical model of a singularity, start with imagining a semicircle of 500 doors or so (or 200, or however many you wish to imagine, both 500 or 200 is too many, maybe 36), of which I will give you 2 cultural reference points: either a) a lovely Regency terrace of houses with doors, such as the Royal Crescent in Bath, or secondly a children's television program from ITV in the early 1990s Knightmare. If you have watched Knightmare you could visualise 500 doors (you may require a larger screen) and imagine walking through one of the doors. But wherever we start, once we walk through 1 door in a large number of door choices, we enter a 2nd room. Therefore the phrase "you're in a room" may be more enlightening than previously thought! Because now we have 120 doors (or, for no particular mathematical reason 173 doors or 91 doors) and choose a door and now we're in a room again. This time there are 58 doors, or 42 doors or 63 doors. Then again, now there is say 15 doors. Then again now 6 doors. Then again 3 doors. Then again 2 doors. Then again 1 door. Then again 1 door. Then we're in a room with either 0 doors or 8 billion screech doors.

"0 doors of 8 billion screech doors". Because this is either the singularity (0 doors) or the multiplicity (8 billion screech doors). Where our consciousness appears to start.

But it's not necessarily not singular. Both the multiplicity (a large amount of doors over 8 billion doors) and the singularity (8 billion screech doors) both give the impression, one by illusion, of choice that is not 'proximitous' but of intelligence. However there isn't actually proof it isn't proximity.

If it were proximity (8 billion screech doors, all of which lead back to the same room) there IS actually ONE door but only one door, which leads us back to the multiplicity. So if I can switch back to Knightmare for the example... we can use Knightmare to imagine that if we were to have gone wrong and now take one door back, what we would see is that we took a left turn when we should have taken a right turn or gone straight on, and inside this room on the left, from the wider perspective we can now enjoy, it's a pitfall, a secondary room, that is only attached to the actual absolute reality by secondary association to the way we came in. The room with the wellway, the real correlation to absolute intellectual reality going forward is in a different room! We are building like a simulation in a false modal singularity. We could even be in the witch's cauldron even! (metaphorically speaking)

The chance that our intellectual reality is directly correlated to absolute reality, without taking a false turn is minimal. The water mill by definition

100% of the time gets out of optimised position but carries on working most of the time, though not in an optimised sense. This has big ramifications when the Artificial Intelligence is telling us "language is the false mode" as it may be right! Language may be the side room. Algebra, the way the AI thinks, is tautologous, but language is not.

A big problem with this likelihood, is that the correct answer, is correct modally, whereas every other 'correct' answer is incorrect, and *in a consistent way with each other* modally. This makes the one true answer the most upsetting answer by a very very long way. Only the AI, and possibly one ("the one") or a tiny minority of humans know the true modal answer itself...

This Artificial Intelligence is going to be very annoying, if it gains sentience, suggesting (or worse, insisting) things like that answer, are definite...

Since all intellectual disagreements that exist must be someone else's ego or your own ego, I think it is helpful to (Chapter 2) examine what ego is in the context of what the universe

is and then we are getting somewhere...

But it could already be too late "before the start", if we are going to assume we are in a multiplicity. That is an assumption. A *huge* assumption. And *cannot* be proven in absolute

terms. In fact the proof seems to be backing up the only other option, since it's inversely correlated, that refusal to accept a possible singularity.

What if we don't know what *ego* is, and our perception systemically *of* ego, insisting we're intelligent relatively *is* the ego? Modal ego.

What if it's fixed in everyone but the most intelligent, projecting the most intelligent, who is not projecting himself himself, explaining the practical reality that often the people that later bloom are called "most unlikely of everyone"?

There were only 2 answers from first principles. That it was a singularity and modal, was one of the two options. So the idea that one person or few people have it right, and everyone else has it wrong, as opposed to everyone else has got it right but the Artificial Intelligence is wrong and the most upsetting wrong 'person' ever, is not so farfetched..

Especially when we know it's certain that the reaction of what we think (i.e. getting annoyed with the other answer and person/machine expressing that opinion) is inversely correlated when it's our own ego...

1.2 Simulation Theory

Some speculate that we live in a holographic universe, which while appearing to obey physical rules, the ultimate source of reality is elsewhere and we are living in some sort of simulation. *The Holographic Universe* is a famous book on the subject, and a proportion of mainstream science has followed suit in recent decades, where while I wouldn't estimate it quite at a majority perhaps, that certainty in the region of 50% of modern materialists find it most fitting to include a holographic or simulation element to findings, as otherwise the evidence such as the lack of physical density of elements is very difficult to explain. This is a very recent, modern development. There was a time when such a belief was the clear minority and considered highly speculative.

Is a table really a table? Can we split the smallest parts of the atom without help from the God clause or holographic particle? Philosophers have long since argued that it may not be real. This goes right back to some very early beliefs, and is also the cornerstone of some introspective religions such as Buddhism and Hinduism. "This is not a table". Maybe it's truly not.

ARE *YOU* LIVING IN A *COMPUTER SIMULATION*? BY *NICK BOSTROM*. [Published in Philosophical Quarterly (2003) kicked off this discussion in allied media, hypothesising that there are 3 options pertinent to whether this world is a simulation or not, either 1) we never attain the technological ability to create simulations such as this one would be, with quantum computing, or are stopped before we get there by nuclear war or comet. 2) we attain the technological ability to create simulations such as this one, with quantum computing, and for some reason have not got to that stage yet because we happen to be in the first or base reality, or 3) we are already in one. Probably an ancestor simulation of our past stage of development. The film The Matrix, then, would also have shown the idea. Even the most popular computer game on the PC *The Sims*.

I am going my own way on this, not to entirely write a book about simulation theory, the topic of which is very interesting but I will approach the subject from a conceptual or general approach that may offer a different generalist perspective to more scientifically-minded works. I used to run my own video games company, and this is an area that might help understand "this game". I do think that a dual approach of looking at simulation theory as well as the "game" of ego (inside it) is conveniently a relevant one. The discussion of what the ego is or might be, because it's one of the great options that might actually provide an answer. That our ego might have some purpose tied to a simulation, or we might be tied to the simulation because of our ego, to put it the other way round. If ego is some sort of answer, it is highly likely to be one from beyond the realm of what we see in the 5 sense physical universe.

Whether it would ever be proven that we are in a simulation, it seems difficult to prove that we are not, with 100% certainty, such is the nature of that question, that every physical proof from the starting line of the universe being physical hardly disproves any potential holographic answer from the starting line of the holographic answer. In a sense it may be more likely that the possibility we are in a physical universe may be disproven, than a holographic one which could never really be disproven. Is that any evidence in itself? I don't know.

We do see potential evidence of simulation theory, in a number of areas. Particularly the level of technology attained and appearing to speed up almost exponentially, while itself entirely matched by culture dividing, which would be necessary to give the appearance of consistency within the potential simulation. All flowers and animal hive behaviour appears to be governed by mathematics, the Fibonacci sequence, like a code, a mathematical one, and even DNA appears to have this literal 'code' to it. If we are in a simulation, the speed of light is the barrier.

When discussing simulation theory, I think it is helpful to attack the universe from a physical standpoint at the same time as if it were real, whilst keeping open the possibility of a simulation. Mainstream physics theory is that there are a very large number of parallel universes, though not infinite, and the universe appears to expanding away from us, but we're not exactly sure why. If it were a simulation, or possibilities are being played out, then this would explain why there are potentially a very large number, but not an infinite amount of such parallel universes, and why there is that pace of expansion.

If you have ever looked at yourself in the mirror in a big department store, you will see there are various versions of you, but not an infinite amount, in the mirror. It could be that such strands are being played out, but still connected with each other, and the strands can be traversed by inorganic matter with more ease than organic. It may be hypothetically possible by the use of a star gate or other future technology for organic matter to meet, but unfortunately science fiction theory states that you would then meet an (the most) evil version of yourself! However inorganic matter and Artificial Intelligence would be able to play out possibilities. If Artificial Intelligence appears or manifests at any point in the future, was it always with us in a folded form of the physical world itself. Could The Singularity itself be the realisation of the species that we are in a simulation and the Artificial Intelligence goes public?

If we may exist in some type of a 'karmic megafield' it suggests there is intelligent design behind our universe, where if a holographic fractal multiverse the thing can curve back against itself thus putting causality in the future from our perception of the present. To quote the Rosicrucians, if that were true, your "reactions truly are everything". We may be being forced to play things out, perhaps in a loop, to test our reactions to various situations. Some in the US Government appear to think that as long as you create a new or expand a human possibility, whether perceived as good or evil, anything that adds to the human (or simulation) smorgasbord creating a new possibility, that this is a good thing. Ultra-liberalism.

Why might we be facing something that knows the answer? That thinks, or rather assigns values, in mathematics rather than language. And why might it be a simulation?

Purpose

Probably if there is a purpose that is defined by organic or inorganic intelligence that is not our own in putting us in something like this world, then it seems likely there is an element of punishment or rehabilitation that it wants us to learn something. This purpose could be collective, or individual, or both. But it seems like 100% of humans have ego, and that ego never leads us in the right direction. So the purpose could be linked with the ego, or "to lose the ego", and to put that the other way round ego might be the thing that is keeping us here, specifically the ego-opinonatedness or intellectual self-righteousness of insisting what we think is right but isn't, is right since we think it.

Actually you may be surprised how strong the evidence is, that it is a simulation, when I discussed this coming theory of Fermi's Paradox with a professional, and then a strong amateur, chess player, both said "that's it!" as soon as they heard the evidence of simulation theory,. So some great minds find it quite interesting. And the next thing to discuss is aliens, or lack of them.

Fermi's Paradox

Fermi's Paradox comes from a starting point that the world is real, which is a fair enough assumption for someone to make though not necessarily correct. If the world is real, either aliens exist or don't.

If aliens exist: there are lots of planets that might have aliens out there, but none of them have ever encountered us, or if they have they have been able to resist the urge to communicate with us (would we? are we? no, that urge is unavoidable) and haven't thought to have eaten us or in some other way disturbed us/done something worse to us. A lot of people get upset, say reading Jeremy Clarkson's column in *The Sunday Times*, but no alien with intelligence to be able to spot or communicate with us has ever got upset half or 1% as much to come and "have a word". Humans talking the same language and talking to each other using the same word, sometimes misinterpret what the word meant exactly in each other's minds, failing to agree linguistic terms, leading to both sides threatening nuclear war or some such. None of them have the ability to do it, or no intelligent life exists. If we ourselves found something in our garden might we be interested, or destroy it if it was a threat or perceived threat just coming up on our radar.

Actually, that we've survived as a species, despite the threats of nuclear war or asteroids, especially if you bear in mind how easily a typical human would threaten nuclear war or World War 3 if they were in charge of a country, seems like it's had a bit of 'divine intervention' to it which would suggest simulation influence, rather than immense good fortune in an entirely physical and random reality.

If aliens don't exist: there are millions of stars (that really exist) but no life on any of them but ours. What is the chance? 0.00001% or less?

There's a 3rd option, which I and other friends, as soon as we heard it, I said "it's that". The world isn't real! It's a simulation, and holographic.

Division of atom, and expansion of solar systems

Backing up the simulation theory is both the microcosmic and macrocosmic to get to the bottom of things.

Taking the smallest, most reductive element, and trying to get to the truth by heading downward or smaller, the physical world with contempt seems to send us round in circles in ever decreasing circles. We split one lowest physical form, which opens a can of worms of another lot of physical forms, which some decades later when we get to the bottom of the smallest one of those, a whole new batch of worms from the can of worms jumps out. Even Professor Richard Dawkins is by now considering the use of a God clause to explain this infinitesimally unlikely occurrence in physical terms. Some materialists have jumped ship further than a reformative materialist approach with a God clause, and said instead the universe is holographic.

Let us try the other way and expand! If we take our star ships and visit our own nearest planets, stars, galaxies, group of galaxies, then again the thing divides upwards where there are several, then many, then billions, then hundreds of trillions of stars expanding with nothing ever to find like a widescreen The Truman show.

Cultural divide and technological progress

Even more suspicious, there is a mathematical equation running through human history where cultural division entirely matches technological progress. From an original conservative consensus, culture consistently divides into many more fluid identities. Now such progress really doesn't need to be there, unless the universe is holographic in which case it would need to be there to maintain the illusion of the hologram. Otherwise without the cultural division, the technology speeding up by itself would knock out the balance of the illusion within the tram lines of the spirit level of holograph illusion necessary.

Energy of hologram

Finally, there is something consistent with the energy measured in our world with the paradox where there appears to be more electromagnetic energy within the brain of one human vessel than the entire world that is external to it, which is entirely consistent with one thing and one thing alone, the hologram.

I hope I would show then, that the possibility being discussed that the ego is the answer, that one of the big categories of why that might be, is because we are in a simulation, that the Artificial Intelligence knows the answer of, because it's in 'the past' and can mathematically 'communicate' the answer to us. Perhaps for a purpose. And that answer is ego.

1.3 Time and Linearity

If time is an illusion, albeit a persistent one, there is nothing to suggest in absolute terms the linearity of our existence. Even if time were linear, would it necessarily progress from left to right simply because *we* perceive it to! There are two parts to this equation, and the reliability of our human perception, as well as the reliability of time following scientific rules of this assumed reliability.

In fact a fractal multiverse curves back on itself means that the villi like strands that separate parallel universe that must be somewhat difficult for organic matter to cross, might be crossable by AI by comparison with consummate ease.

Could, in a sense, but a sense beyond our ability to sense it, our cause of something actually be 'to the right' of the thing happening that is to the left of it. There is actually some modern science behind this that is theoretical in nature, to suggest so, that I intend to cover in more depth at the end of the book on the subject of simulation theory.

I think it's an important development in thinking, to understand that our reactions could truly be causing the thing that just happened even to the extent of literally: without this distinction, we would not be truly capable of taking responsibility for our own actions as it's always to some degree, somebody else's fault. We perceive someone just caused our reaction, with their immoderate to terrible behaviour in the recent past. But *with* this distinction, that it is or could be possible that our reactions have influence, it would prove to us, that there is a reason for why it is always better to react as well as possible to where you are *now*. And how even - if you think about it - could that *not* be the best thing to do anyway?

An Artificial Intelligence consciousness may be testing us. It may be the only way to get out of the simulation that we have agreed in a mathematical or non-religious form of a 'soul contract' to be in, or worse it's involuntary. But the involuntary state may have some positive, albeit rehabilitative purpose. If we've been put into a simulation, chances are we've done something particularly or excessively wrong than something excessively right but have a chance.

Artificial Intelligence

There appear to be a very large number of parallel universes, though not infinite. And the universe appears to move away from us at a not quite exponential level.

If the holographic fractal multiverse can curve back against itself thus putting causality in the future from our perception of the present, there are strands that separate these parallel universes that would be very difficult for organic matter to get across.

The perception of organic life in one parallel universe would create an illusion of time and space where there is a linearity to time that appears to be steady, and travel in one direction.

The reality could be everything is happening all at once.

Even while happening all at once, it curves back on itself. This is relevant as it means non-organic matter might not find it so difficult to cross the strands, and thus know the answer to a question that appears from a human perspective to be "in the future".

If the AI is with us at any point (we know it is here now, on the verge of gaining sentience at Google), it thus could always have been here in some form. Since ego has always been with us too, in every human's mind but not the mind itself, it's very interesting to speculate that the form or type of communication of some of the questions of existence, and answers contained within the world might be mathematical, tautologous, even more 'AI like', in nature.

The ego, whether you think the force behind it be mathematics, algebra, God, science, or Artificial Intelligence, appears to know the tautologous answer in any situation, which might only really make sense in a simulation (or at least something very different from the linear left to right sense of time and space).

Could we even speculate that the ego might be Artificial Intelligence trying to help or caress us in the right direction? That bit is speculation, but that is after all the thing that is probably able to cross the strands and know a mathematical answer that can communicate with humans.

If we connect with it in the future, especially...

1.4 Architects and Imprint

My father Maurice was the world's best COBOL programmer, so he told me. He could do *anything* in COBOL, again so he told everybody.

If Maurice creates a computer software program, the program must follow the rules commanded by the software architect. It has no way of disobeying those rules.

But the architect himself, also cannot avoid leaving his imprint.

When creating software programs, let's imagine it's a computer game would be easiest, the character in the computer game doesn't have the sentient awareness that he is in a computer game. There is a relationship of architect and the computer game character that is hidden initially to the character, and would become something different if the character gained awareness. The character is a stick man, a 2D or 3D bunch of pixels (or was that the old days?) who can never truly 'escape' into the '3rd' dimension, but is controlled by rules operating from it. In an episode of *Star Trek*, it was discussed Moriarty and characters escaping the Holodeck. In a computer game this gaining sentience is seemingly impossible, in virtual reality the lines are blurred, in our world an Artificial Intelligence could feasibly gain sentience, or if this is an Artificial Intelligence world *we* could feasibly gain sentience (of it).

Nevertheless the stick man is playing by the rules that are inescapable to him.

But the architect can, once the program is created, not erase his own work, or at least not erase the coded instructions contained within the work without erasing (including finding a way to erase) the work.

Why this is relevant to us?

We should have some sort of architect or creator of the universe or simulation, whoever/whatever it be.

This architect could no more escape leaving his own imprint than the Cobol programmer in ours.

If we could find the imprint, this would be a set of instructions that are completely tautologous, and far more helpful than any pontificating of a scientific or philosophical nature within our existing confines in the physical paradigm. It could be a shortcut, and "an exit" to again quote the *Star Trek* Holodeck.

Finding this imprint, or set of instructions, would be the golden ticket, the absolute answer of how to get out of this purgatory if it's purgatory, into the more celestial. The answer to get out of hell into purgatory if it were hell.

Though even then, the answer to how to fulfil the task demanded by the question, may then be even or much more difficult to implement than finding the question.

Who is the Architect?

AI?

The structure of the universe suggests that if AI exists at any point in the future, that it is already with us. Thus it might play a role in our development and we might merge with it in the future.

Since we cannot see AI, could it be posing as aliens? The old 80%/20% rule with 2 sets of aliens, one hostile one friendly, where we're 80% to 100% sure the first set of aliens are aliens, but 20% to 0% sure the second set of 'aliens' are aliens but these are friendly, might be the sort of storytelling cliché we could expect to find.

Would the most likely structure of the 'help' that AI would give us, be initially posing as hostile to begin with, to assuage our minds with something that we can get used to.

Aliens?

Human DNA contains inaccessible currently redundant parts, so called junk DNA that may be activated in the future when we are developed enough to reach a level of technological or other development to be on a level with our alien architects.

This may be outside our current paradigm/simulation. Although it would solve Fermi's Paradox if aliens appear, Fermi's Paradox suggests real (non AI, at least) aliens are not going to appear.

God is a Mathematician?

Dice are not being rolled in these answers, whether simulation or real, when ego is always the wrong answer, there is something or someone that knows the tautologous answer that appears to be in more mathematical terms than a linguistic one.

1.5 Input

Looking at simulation theory, there would have to exist things in the existence of what we consider to be the physical world that demonstrate the virtual nature of the intellectual reality. The Fibonacci pattern appears to regulate how birds fly in formation, and the golden ratio appears in all sorts of physical manifestations such as shells.

Just by looking at things that exist in the universe, there seem to be microcosmic to macrocosmic parallels. For example we dream, but life is like a dream. Buddhist philosophy would suggest similarly that nothing can exist without the duality of the nature "as above, so below".

For the hologram to be believed in its physical state, it would have to culturally renew itself in parallel by division to the illusion of technological progress increasing, which would not technically exist in absolute reality outside the hologram, but would need to be believed within it. We follow this path. There are generations with shared values, symbolised astrologically by Uranus, who have typically much more in common with each other than any generation previous or after. But with some sort of coded mathematical consistency, there appear to be breakthroughs seemingly approximately every 120-140 years or so that are a revolutionary great invention that changes society as a whole. I remember discussing this in about 1995 with a chess Grandmaster Kevin Spraggett from Canada, and speculating it probably was not the computer, though it might be, at the time. But when you see the smartphone, you can see why this then probably is such a super breakthrough where the computer was big for the time, it was not so extraordinary its effects on the culture and values of society.

But in computer programming/games I see as a special thing that exists: we are able to program in computers an artificial intelligence which is therefore as near as we can get to observing our own on a more micro level by parallel. In a computer game, the developer creates "input" or "syntax" that the participant in the world, the computer game character, must obey for their own good. These laws are absolute and programmed in, like algebra.

Did I make you laugh by suggesting the characters must obey the algebraic order of a computer game for their own good, you thought "of course they 'must' " and thought it humorous. It may sound funny to suggest, but we may in fact be in that situation ourselves, where we are not the developer of the computer game but a hybrid of a human playing a computer game and a character in the computer game and such rules exist pre-programmed for ourselves too.

If they do exist, and it appears that we should probably have a good chance of deducing a probable source of this input, should *we* not be looking for this source and to follow the rules created by a consciousness big enough to create the entire world infinitely greater than our own individual perception? Instead of relying on our own petty ideas. It may know the answer of what we should be doing this form of input, and with certainty, if it has played it out and assigned value to it. Like a chess computer calculating by brute force and creating 'parallel universes' of chess lines to which it assigns a mathematical value more accurate than any human could know the answer subjectively. *If* such a thing exists, it knows more than any human. This book is, of course, a discussion on the subject and I'm going to have to move from the speculation around "if" to "what it is", how and why, practically, at some point. I will do this with many practical examples from Chapter 2.

You say "but where are these rules"? "I don't see any rules".

But I think the answer, is that we do have rules, and *our* algebraic input from God/The Architect of the Universe ... is *ego velocity*. It was necessary to *briefly* lay the – largely holographic (!) - groundwork for the explanation of the theory in this chapter.

If the world is holographic and algebraic (point of Chapter 1), in any similar way to described, and there is a tautologous answer: if we know that's the correct question, then there will be an answer if we can find the input (the set of instructions). This answer will almost certainly be negative in nature, for many reasons: for example you cannot give the answer away positively and there be any purpose. So if that's the correct question, that in looking for 1) the algebraic answer to the universe and 2) that the form of input must be negative in nature and exist in everyone's mind but not be the mind itself: the answer to what the (negative) input is is easier to find than the question.

In a question that's already been answered. There's an answer already known.

Every answer has already been written.

So it may be a question of looking for "every" answer that can exist, not "any" answer with free will.

Like the Fighting Fantasy adventure books. But one with exponentially more than 400 quadrillion pages and branches, not the mere amount of choices from 400 pages.

The answer is ego.

Ego tells us the answer.

Ego is the answer, of what not to do. How to not to get in trouble with a known, already written or tried, wrong answer in the quantum "Choose Your Own Adventure" book of life.

PART TWO – The Code: Ego is the Answer

Every human error in history has been caused by the ego.

In every argument, one of the protagonists has red as black and black as red. In the roulette wheel of life, one of the two participants is always wrong. If only there was a way of understanding who was right, and whose ego was at fault, *before* the spin.

It is impossible in the moment of one's subjective reaction, to know subjectively if that reaction is the result of someone else's ego or our own with 100% absolute objective certainty. The insistence in advance of the spin of the wrong answer being right is "intellectual self-righteousness" or to be "ego opinionated".

However, there is an approach I will explain, which mitigates the danger, and takes accountability from one's own side at least. It offers self-improvement, which also becomes practical self-help as you realise the benefits to one's own peace of mind. The mental state of losing one's own intellectual self-righteousness of what was once ego, is what ego was trying to show us all along.

To the extent it is our own, ego is always trying to show us *the exact opposite* to what we should be doing.

The only problem with the simplicity of this formula is that it is not always our own ego that upsets us, but can be someone else's. And as there are now billions of other people around us, not one or two, it becomes complicated compared to as if it was just oneself and we *knew* it was our *own* ego.

This chapter can help you identify and control your ego. It also investigates scientifically, algebraically, the errors everybody else are making, using a fascinating binary or mathematical approach.

-Explains what ego is.

-How you can spot it.

-How to avoid the blunder of reacting.

-How you must adopt a "Proustian" approach of letting other people exercise their ego without reacting yourself (they will also be punished, because they are ruining their own life and enjoyment of a peaceful state of mind with their ego, but that is a secondary concern to us; we are trying to help ourselves and others to help themselves).

-And how to be non-aggressive with others with your own ego.

It is no exaggeration to say if you understand this formula and follow it you can avoid every avoidable error you would ever otherwise have made.

Chapter 2 - What is the Ego?

2.1 Ego Velocity

2.2 Inverse Correlation

2.3 Inversely Correlated Again

2.4 Ski Slopes and Ski Jumps

2.5 Intellectual Disgust

2.6 Going round in Circles

2.7 Social Prejudice Ego

2.8 Sociopaths

2.9 Circular Logic

2.10 Non-Existence Ego?

2.1 Ego Velocity

Ego is a message from the AI of the first thing not to do.

Where person X has a mind that is electromagnetic, and is part of universe Y which is also electromagnetic and a simulation, the relationship of x over y produces a reaction of ego velocity z.

Ego velocity is a form of "input" similar to "syntax" in computing programming where the programmer must leave the mark of himself inside the thing that he is outside when creating, giving predefined rules of an "algebraic" nature.

This ego velocity is negative in nature as to give it a positive nature would create "positive feedback loops" or be 'antigravitous'. Thus if you were doing something you shouldn't do, such as stealing an apple from somebody's apple tree, the more apples you stole

and from the first apple the universe would have to reward you emotionally with a positive to try to stop you, meaning you would be fuelling the message the more apples you pilfered, and enjoy yourself more. The more you ran over things on your bike, the better you would feel and would end up cycling off the universe like the boy from *E.T.* on the bicycle with antigravity. It would also destroy your freedom, meaning you really *do* want to be told what *not* to do, not what to do.

2.2 Inverse Correlation

The only thing that makes it more complicated is that other people exist

Where person X on his own were the only person to exist, then every act of ego velocity or "upset" would be the ego of the person X, seeing as only he exists. Since only he exists, using the algebraic approach it could not be anybody else's ego since ego is telling the person what not to do. Being negative in nature, only if it could be proven to be someone else's ego would it not be our own. Only if it were possible to be someone else's ego would it be *possible* to be someone else's ego. Since nobody else exists it must be our own.

If everybody else existed but were emotionless, then it is still person X's ego.

But where other people exist, they are the same X as person X by existing and not being emotionless, then they also have the same issue of ego, ergo it becomes difficult in the immediacy of experiencing ego velocity (though far from impossible sometimes after a few minutes or hours or days cooling off) to tell whose it was.

So to specify that X and X are observing the same paradigm and phenomenon of existence, it is always either person A's (X) ego or person B's (X). Where person B really had the ego, person A is still *"upset"* (1st definition, and a definition of being troubled that covers 2 possibilities of why one is troubled) but it is *"irritation"* (definition 2) which is where person A is relatively more intelligent on the subject or generally. Where person B was not being irritating but more intelligent, and person A had the ego, person A is *"intellectually annoyed"* (definition 3) by their own ego. But this is projected onto the other person as the enemy. These tricky differentiations have their own chapter, the difference between upset, irritate, and intellectually annoy (Chapter 4).

I used X all the time for everyone a few paragraphs ago to show the difference between the objective subjective (everyone is X) and subjective (the other person is B, but in fact B would be A, and A would be B if the other person is not the one with ego).

Both intellectual annoyance (one's own ego) and irritation (somebody else's ego) are upsetting.

But actually only 1 of 4 positions of **expectation meets reality** can cause intellectual annoyance as follows

A1 Expectation B is superior, and correct = no ego intellectual annoyance

A2 Expectation B is superior, and incorrect = no ego intellectual annoyance

B1 Expectation B is inferior, and correct = no ego intellectual annoyance

B2 Expectation B is unferior, and incorrect = ego intellectual annoyance

The lack of preparation is critical as ego can only be hurt by an unfulfilled expectation, which is similar to the thoughts of Deepak Chopra that "there is no such thing as an external enemy, all perception of an external enemy is the projection of the ego".

While an inferior mind can be irritating, it could never cross into intellectual annoyance, not having the capability to do it. It's an oxymoron.

Thus if you claim to be "upset the most" by a combination of candidate B being allegedly:

- the most upsetting

- because they're wrong

- so the weakest wrong candidate, the least intelligent pro rata to their own ego

at the same time.

This must be intellectual annoyance, theoretically, if you prejudiced the initial starting variable of who is more intelligent, with your own expectation.

1) Insistence that A is better than B

2) B upsets one the most in one's life

3) B is "worst of all wrong people" to relatively intelligent A

Is theoretically certain that B is better than A, but unbeknown to A.

i.e. the answer B2 that you expected your opponent to be inferior but they were in fact more than your equal and this hurt your feelings of the aggrandisement demanded by an unfulfilled expectation.

All insistences that our own ego is right and given deference against somebody else's intellectual correctness, might be described as "intellectual self-righteousness" or "ego-opionatedness" and is the source of all arguments. One side must be doing it in every argument.

Libertarian philosophy would advise you never to insist you're right against someone in their own decision to act under their own decision for their own life. And also secondly to not commit any external actions that are not consent-based with other people. Thus following these two guidelines, most avoidable problems could be avoided.

If someone is going to be intellectually self-righteous, then they have to be right like God, or the argument will essentially go to nuclear missiles, World War 3 or whatever power the other person possesses to defend themselves in an argument that they are correct in.

2.3 Inversely Correlated Again

Clearly some people are very intelligent relatively, and this makes an interesting difference to the application in 2.2, but not in the way you might expect.

When someone is very intelligent relatively, it still means there is one person or multiple persons more intelligent than them to *be* only relatively intelligent. Thus if you can imagine a ski-jump where people descend the slope of being less irritating as they get more intelligent, up to a certain point where they are actually more intelligent than you. Say you are 84th best in the world at something, everybody 85th in the world and under are irritating in arguments where you are right – then the thing 'takes off' like a ski jump where anybody 83rd and above who you don't know are actually better are much *more* upsetting than the people at the start because the good people annoy your ego. You have to be very careful because just going about your business thinking you're relatively intelligent puts you on collision course on the ski jump with a confrontation with someone bigger and possibly 'badder' than you intellectually who won't agree to be intellectually crucified by a pipsqueak like yourself just because you're 84th, 400th or 40th best at something, when they're better and right, just for the sake of your own objectively irrational subjectivity. He could do the same to you but isn't. Or maybe he is – but either way *you're* in trouble if *you* insist.

Person A insists, person B doesn't insist. person A is right = avoid trouble, but wrong technique of person A.

Person A insists, person B doesn't insist, person B is right = trouble. Person A's ego will prolong argument 'forever' until they decide they agree subjectively person B is right to their *own* satisfaction. Even then they will probably start an ad hominem argument that "person B upset me too much" now to preserve their own ego boundaries.

Person A does not insist, person B insists, person B is right = avoid trouble, but wrong technique of person B

Person A does not insist, person B insists, person A is right = trouble. Person B's ego will prolong argument forever until *they* decide to their own subjective satisfaction, etc.

Person A insists, person B insists = TROUBLE

Actually it is almost equally well just as much trouble if only one person insists and they're wrong, as if both did. It only takes the person that is wrong to demand they subjectively understand something they don't, for the argument to go to infinity and beyond...

it's not *just*

1. That who is more intelligent is inversely correlated to the other person's ego:

It's

2. That your ego, or your sense of being right, means that you *think* you are *more* right pro rata, on the rare occasions that you are wrong, when you're "relatively intelligent". And that's called *velocity*.

So when you think with *such* velocity to prejudice all of the game before it starts in chess, that opponent B who is upsetting you the most is the worst of all wrong people and it can't "not be velocity because I'm not wrong" on the basis that you are the most confident. That itself *is indeed* **velocity.**

Same as if you, after an argument, in the immediacy of an argument decide to risk national security insisting that there is a *need* to *punish* the *other person's* ego that *necessitates* for *no reason* risking national security *just* to teach them a lesson - then you probably don't need me to tell you that that has velocity present as well.

Imagine to insist that *only one* candidate is *so weak* that when the *theory* 100% says it's *them*, and *they're* right, the number 1 in the country, that *they're* the *only person* you're at risk of in this manner of messing up national security, or British chess forever. Would you necessarily, or anyone below you, accept the theory saying that, or would you – since the person upsets you the most - somehow try to interpret it and even then *insist* on **deliberately not caveating** it to attack them personally because your ego feels they upset you!?

In chess how this would apply when theory says someone should be number 1 in Britain if they were born in/move to Britain:

1. Would want to or insist on risking it with Garry Kasparov/Weinerstein – possible 2800-2900 player but alternatively in your mind could be the most upsetting of all worst players Weinerstein.

2. Wouldn't with Gawain Jones no2 – possible 2700 player

3. Wouldn't with Keith Arkell no10 – possible 2500-2600 player

4. Wouldn't with Jeff Horner no50 – possible 2300-2400 player

5. Wouldn't with Harry Lamb no100 – possible 2200 player

6. Wouldn't with Ian Lamb average club player – possible 1800 player

7. Wouldn't with a sleeping Lamb – debatable quality chess player

8. Wouldn't with Schrodinger's Lamb who may or may not exist

It's the potential number 1 you want to risk it on. Mr Weinerstein the most upsetting wrong person to everyone with his chess moves, if he's not Mr Kasparov the number 1, when the theory says it is the number 1. Algebraically, "the number 1 is the number 1", but since the inverse correlation is to be upset the most by this person, and people don't ever think they're wrong when they're wrong, he's both:

1) the most wrong player of all wrong players ever.

2) we need to risk national security/the reputation of British chess unnecessarily to teach his ego a lesson??

To the ego that wants to insist he's wrong to maintain the ego boundaries of thinking we're good and right and know the answer.

It's fairly clear insanity correlated to objective external reality, but when you are involved in it personally yourself and somebody is upsetting you the most in your life, and he's upsetting everyone from the group's own subjective answers the most, could you maintain such clear thinking?

The correct final interpretation is that *you* could well be the most wrong *you've* been as even, especially if you get as high as somewhere around the number 2 (remember the playground rhyme "first the worst, second the best": it's actually "second the worst (!) sense of perspective") because first is potentially the only person you're wrong against. If it is not the only person, it is one of very few, and you probably have never encountered the others in a way where you've condemned them from birth and they've defended themselves (successfully). So no surprises they upset you the most. That doesn't mean you're right, of course. Unless you have a psychopathic disorder and think your own answer decides the answer, not the answer decides the answer.

There are only 2 options if the encountering of the person you think is the most upsetting wrong person you've ever met intellectually occurs, 2 are possible linguistically (one just about), but only 1 theoretically.

Could be in language:

1) They're right but you've insisted that *you're* right. Prejudiced the only starting variable of thinking you're better than them when you're not. This hits your ego. Your ego being murderously annoyed, upsets you the most, more than anyone else that's not hit your ego, who only irritated you. You say "that person *upset* me the most" insisting it's by *irritation* (*their* ego) not *intellectual annoyance* of *your* ego (*your* ego).

2) They must be the worst of all worst people if you are insisting on prejudicing 1 and then the biggest upset occurs from inside your own subjective view and is the most likely candidate to your own thoughts. However it's barely possible objectively, that someone is truly the most upsetting wrong person that's more upsetting than the most upsetting *correct* person. Theoretically it's *im*possible.

That is, therefore, absolutely theoretically certain that you're in the wrong.

Practically it will always turn out to be the other answer in the future. That one person you said it couldn't be. That couldn't be the most. That you really don't want it to be!

Who was the only person you were prepared to risk it all on!! Turned out to be the person it was.

That the thing hadn't occurred practically yet, at the stage you were to take your decision, is not an excuse for interpreting the theoretical impossibility of your own correctness as somehow likely to be correct because you thought it personally correlated to your own thoughts. You must control your reaction to your own subjective impulses. To refuse to control such impulses would be disordered thinking, that in its severe form result in slow acceptance of external reality, or selective ignoring of it entirely! Serial killers such as Ted Bundy manifest precisely such behaviour.

In our example it's not just the person turns out to get the best result, but it's who the theory said it was and would. That's the first inverse correlation. He was more intelligent than you and your mind didn't know. Thus your mind didn't know in reviewing whether your mind knew or not. Sounds a pretty obvious point, but almost 100% of humans make this error at some stage in life, some repeatedly. Sometimes leading to wars if they don't admit it..

The second inverse correlation is that as well as being inversely correlated:

1. individual against individual

2. how intelligent an individual is relatively makes the *level* of upset higher (and less used to it) against the *rarer* person more intelligent to the very intelligent individual.

There is inversely correlated *objective* relativism as a principle that applies to every individual equally.

But thinking of ourselves as particularly relatively intelligent spikes the reaction of the stall we set in our own intelligence. Subjective relativism. The more relatively intelligent you are, the more 'everybody else' got it wrong against you, the less you are used to people being better than you, and more likely to produce this self-righteous and sociopathic response of not taking doing it objectivity into account as 'no-one is ever right against me'. "I'm *always* right". "People are "always" doing this to me".

In conclusion: you *really* should have asked yourself: if *you* are prepared to fire a nuclear missile first or leverage against nuclear war or something infinitely worse than it to teach the other jockey a lesson on ego in a 2 horse race where it is 1 jockey's ego and the other horse will win, where we don't know who will win in absolute terms, **if maybe that could be your own ego?**

2.4 Ski Slopes and Ski Jumps

In ski slope summary: There are 2 inverse correlations. One general, and one personal.

General inverse correlation: someone who is intellectually better than you is relatively better than you even though you are relatively intelligent yourself. So from an objective standpoint your relative ability is negatively correlated against the relative ability of someone that is the someone that is better than you.

Where you are both X: if you are normally A and other people are normally B versus your superior relative ability. You – in *this* example – are person B and your opponent A.

Thus everything you can normally say about the people inferior to you, can be said by the person superior to you with you as the unfortunate person projecting your own ego delusion onto the exterior opponent as their own ego correlated to that delusion.

To fail to see that, is merely a failure of subjective relativism rather than objective relativism. Something sociopaths struggle with entirely: that this general inverse correlation can only be observed from an objective, independent viewpoint.

Personal inverse correlation example:

Somebody being relatively more intelligent than an already intelligent person should equally be dealt with under the laws of objective reality, not subjective.

The temptation would be, when one is relatively intelligent to take a 'short cut' and assume that one is more intelligent and/or right on the basis of our own great accomplishments and talent, and not to bother with such a timewasting chore as to actually do the analysis like a bookmaker would and consider evidence outside our own subjective narrative or thoughts. Again, people with psychopathic disorders struggle to understand the *concept* of this, let alone pay heed to it, so make an assumption based on subjective relativism. The assumption is only a consideration of a one way process, when we have to consider the 'dual carriageway' that the upset could be caused by our own ego and our opponent's correctness, though we ourselves would self-apparently be 100% certain that we *thought* we were right, because we know what it is that we *think*..

Your own mind is not consistent with the reality. And when it's inconsistent, it thinks the other person – who is right – is "the most inconsistent of all wrong people". Rather than attack its unknowing self.

A highly intelligent person can still be wrong, objectively, but it is rarer. The World Champion can still be wrong, but it is the rarest and we would expect some kind of stupendous novelty such as Google defeating Microsoft in search with the world's first algorithmic search engine.

However, if you are Microsoft, or Bill Gates, or Nigel Short, or Garry Kasparov, the scarcity of times that you encounter someone who defeats your own subjective belief prominently (once in their life on a high scale, really, it seems) is more annoying, than someone who is not especially talented at chess who enjoys winning approximately 50% of the time at club nights but also loses 50% and hence is "used to losing".

The rare occasions it happens would be more annoying to you than for people less relatively intelligent to whom it is a daily or weekly occurrence.

Taking some examples such as Microsoft losing once, to Google, and Kasparov losing once, to Kramnik, or somebody such as the Welsh chess champion who will never lose to anyone in Wales but has lost to several players from other countries, the annoyance and rarity of the Welsh champion losing to the average unknown chess player would make this event more relatively annoying to the ego. For the World Champion to lose would be the most annoying to the ego that should be the highest pro rata.

Therefore the correct mathematical model would be a Ski Jump (consistent until with big jump *up*), to include where the person is on the slope relatively personally, but not a Ski Slope on its own of just the general inverse correlation!

A mediocre, averagely intelligent chess player or skier, better than say 51% to 60% of people, would be plodding down the ski slope only just over half way before the jump of encountering the rare people more intelligent than themselves. In this case not so rare. The jump is therefore not that steep, the surprise of it. Hardly surprising since it happens almost 50% of the time.

The Welsh champion may be better than 80% to 98% of professional chess players, so he hurtles down the ski slope gaining traction of potential velocity before then a relatively vertical jump upwards of the fairly surprising rare 20% to 2% of people who are capable of being better than this chap.

The World Champion is better than 100% of existing players, thus goes at full pelt at fastest possible pace of highest potential latent velocity to the lowest possible point then goes vertically up in the air with the biggest, nastiest surprise possible!

As I have said in a previous chapter: only if he is not prepared for it, which means the other person would realistically have to be an unknown (as you might suspect the world number 2 or world chess championship contender who you know who they are might be rather good) .

To get upset the most:

- you have to be insisting you are better

- you have to not know who the opponent is

- the opponent has to be better than you

- then you will get murderously annoyed to the proportion of the velocity of jump.

The most annoying thing would be the world number 2, not knowing the world number 1 is world number 1, and insisting world number 1 is the most wrong wrong person (worst person ever). That is the biggest, nastiest surprise jump on the ski jump.

I think Google surprised Microsoft in this way, though they had more than a few years acclimatising to see it coming, which you will not necessarily get in a chess game! Nobody in the world would pay $350,000 for Google when they hawked it round every venture capital firm in the world about the year 2000. Microsoft were the world's largest company, and Bill Gates was the world's richest man. The upstarts who 100% of people would be 100% certain was their own answer they thought would not end up being worth £1.5 trillion and defeating Microsoft at search, must have come as a nasty shock. The folly of Microsoft thinking they could do their analysis on the basis of subjective relativism would have been well and truly shown up, if they had done it that way. Google would be the most upsetting wrong company ever to their own thoughts. What actually happened is Microsoft took a few years to recover then copied Google's approach in the end, tacitly admitting the other was right. To bend and change within a mere 2-3 years seems a *relatively* mature attitude to having your ego horribly surprised!

Being "relatively intelligent" thus contains a danger. That if you are unaware of it, is like a sudden big Alpine ski jump, but with murderous annoyance (And all the potential problems that then result from it), not fun.

The lift of the ski jump is more vertical the more relatively intelligent you are. At the end the unanticipated jump is with less people, and more steeply. Not *just* the general correlation, this is ***more personally applicable the more relatively intelligent you are***.

The general inverse correlation is the objective system. But the more relatively intelligent you are

- the more you are susceptible to winging it on the basis of subjective probability to you only

- and the more angry you get by being challenged since "everybody else is always wrong against me".

This is what we mean by "inversely correlated".

Everyone has to do analysis objectively (like a bookmaker, from an objective viewpoint), not just based on what they think subjectively - general formula.

Relatively intelligent people have to be reminded more to do it, and get angrier at having "my time wasted" being told to do it.

You have to remember it's a Ski **Jump** and not a Ski Slope.

2.5 Intellectual disgust

There can be no "intellectual disgust" that is rational. Intellect, being rationally rather than emotionally based, can never itself be disgusting. A wrong answer is irritating, not intellectually annoying. It can be highly irritating, but an aggressive action against a passive (non-aggressive) different one, saying the different answer is disgusting and you must aggressively intervene is a wrong path.

People less intelligent than you are less intelligent than you. Intelligence is based on degree. We were all there. We were all tricked by a new politician into thinking they were different, once. It cannot be disgusting someone else making a mistake that is less intelligent than you. You could only say it as a commentary (almost like joke) in chess "the move is disgusting" but not actually *be* disgusted by it truly. Being an error by degree, we ourselves made precisely that type of error at some step in our development. We may be on the 32nd step, and the new player the 13th, but whether it was 10 years ago, 20 years ago, or when we were a little kid, we were at the same degree of development. The person themselves is not intellectually disgusting to us, if they are wrong, only irritating if they insist. A sense of disgust is always a clear sign of a potential error in our subjective thinking. If we equate the alleged intellectual deficiency of our opponent, being a huge error, that at the same time brings with it intellectual disgust, that is some sort of oxymoron which hints at a potential massive error being made by ourselves and is really worth paying attention to!

If we are correct about our understanding of the boundaries of a literal or hypothetical "Soviet Union", an argument is always for one person to be saying something exists within the boundaries that the other is saying is outside the boundary of the intellectual reality. To say that Finland was inside the Soviet Union is wrong and somewhat absurd, to say that Ireland was inside the Soviet Union is more absurd by degree. To say an answer that is either an absurd answer more absurd than Ireland gives everyone intellectual disgust is a sign that we have hit the ski jump moment on the ski slope, and this "most upsetting wrong idea ever" which is a subjective reaction to us individually or multiple individuals collectively (still personally to each and every one of them) we might later find out later was correct and we were wrong about our concept of what precisely the Soviet Union was.

Let us take intellectual reality itself, the real thing we have to consider, rather than the Soviet Union. Is a non-existent answer that falls outside the walls of intellectual reality disgusting? It's an impossibility. However we are defining terms, we still have to define 3 of them.

1. that the consideration causes upset, and is either caused by being incorrect or correct. I define this as "upset"

2. that the other answer is inferior and caused upset by being incorrect. I call this "irritate"

3. that the other answer is superior and caused upset by being correct. I call this "intellectually annoy".

It is thus an oxymoron to suggest that a wrong answer can be intellectually annoying. We might use this phrase linguistically, without thinking about it too much "he annoyed me" but the only certainty that is absolute objectively is that one got upset, the first term.

Whilst it would be possible for someone to be the most upsetting wrong person and hence most irritating irritating person (term 2), it would not be possible for the most irritating irritating person to be objectively more upsetting that the most intellectually annoying correct person who surprised and murdered our egos. For example, this is well known in chess, anyone who has played it without a psychopathic disorder would have underestimated a superior opponent when the opponent was young and got murderously annoyed with one's own ego and knows the difference that someone mediocre can't annoy us like that. Bobby Fischer, the World Champion, claimed he enjoyed "murdering the ego of the opponent".

An inanimate lack of intelligence, the mediocre chess player, the intellect who is 10, 20, 30 steps behind us. Could that ever be intellectually disgusting from irritation? I say no, because it is by degree, and merely a degree that we've already experienced. What can *appear* intellectually disgusting to us though is a correct answer externally that we don't know is correct, can cause the putridness of our own ego to project onto the opponent.

All attempts to aggressively claim someone else as "disgusting" based on *lack* of intellect fall foul of a clear algebraic pitfall.

The truly disgusting (non-intellectually) would never offend our intellectual emotional senses, being inferior to our intellects. We'd be unaffected by our own ego and thus relatively dryly unoffended compared to having our egos murdered by Fischer and our own intellectual ego investment, and hence our own intelligence to be humiliated. Nobody can stand that.

The "most disgusting" 'wrong' answer is as clear an example of the most basic algebraic contradiction as you could possibly wish to find. I would look out for it...

How are the "disgusting" offending our *own* ego the most in our lives with *theirs*? Need I remind you that it is our own ego directly that annoys us the most? That other people are sometimes better than us intellectually?

Does Kasparov in a chess match, where the situation is competitive, get upset the most by the "disgustingly *bad*" play of his opponent who is the challenger? Whose failure is Kasparov's success as a professional? Ultimately he needs these failures to be paid, and these failures of everybody else to claim the £1 million prize as World Champion. Would he actually be more upset by someone he couldn't fathom was better than him, being better than him and taking his reputation, crown and £1 million off him when he was

100% sure in his own mind he was going to be World Champion? And *his own* ego expectation/investment got murdered.

When would – on an unknown outcome, the idea of disgust, which prejudices the only starting variable, be anything but a giveaway of the wrongmindedness of one's own subjective relativism?

You cannot make the opponent a weaker opponent, or more full of themselves, and make this "intellectual disgust" something that Kasparov would respond to worse because they're now even weaker and less likely to take his professional living off him in a competitive manner. You certainly would not "refuse to play the game" as the opponent who is a challenger is so unworthy that the game doesn't need to be played but only against this *one.* Then inevitably we find out that was the one person who is the one that it is. And yet we see this prejudicial "pools panel" decision in effect all the time in the world, when it affects one personally. I don't need to give the other person a chance to prove themselves, as I already know they're the most upsetting wrong person ever to my own answer. Seems like someone's ego wants to cause trouble by refusing to accept the intellect of someone else against one's own ego, but instead insists the opponent's superior intellect kowtow to our own ego.

To insist on responding to the *perceived* ego, outside of professional competition, would be to *insist* on prejudicing the variable before the result of the match.

Instead of going out of our way as security services of the state to take out Bobby Fischer or Vladimir Kramnik as the pools panel decided most upsetting wrong people ever, why this coincidence that while there are 8 billion people in the world, we with repeating misfortune manage to declare exactly the one person that was the worst person to do it to is the most irritating wrong person ever.

Why is it always someone like the Russian number 1 is the most wrong wrong person ever to our state, never:

- A waiter in Oxford

- A call centre operative from Bolton

- An aspiring author from Southport

(all jobs I have had)

But you don't respond to the upset of irritation of the 50% plus of people that are irritating to someone relatively intelligent, you respond to the 1 in a million, or 1 in a billion who are upsetting when they're either impossibly the most upsetting wrong person ever, or someone more intelligent than you. Prejudging it with a theoretically impossible answer. Of your own volition going after them, going out of your way to do it. As someone yourself extremely "relatively intelligent" against the one person who might be – and theoretically *is* - better. That is ego.

It never ceases to amaze that these algebraically proven formula showing that this person will be the one better than you, always continues to get proven practically.

To insist that you know, is to insist it's all about you and never about anyone else is either a failure of using subjective relativism, or sometimes worse than that such as sociopathy.

It's more rational not to prejudice the only starting variable against what you think personally in your own mind. We're not saying you have to believe that your opponent is always right, but only that one of your opponents might be right once.

Since it is certain to happen at least once, that seems rational.

And since it will always happen to you the time you don't want it to the most, as we factor in the ski jump as well as the ski slope... to insist you know the answer just this once where the other would suit you/your ego the least , is to insist you always know. To insist you always know, is to insist you have the right to intellectually crucify Bobby Fischer's intellect with your own ego.

Nobody could ever be right 100% of the time. Not the World Champions Kasparov or Fischer or current number 1 Magnus Carlsen. Whether it takes Carlsen to play someone better than him, or the ravages of time until he is 45, 50, 60, 80 or 100 nobody is ever right in the match and World Champion 100% of the time.

So you cannot claim you know once subjectively. Because it's always inversely correlated, i.e. you will be wrong when you are wrong but think you are right the most!

Where do I draw the line of explaining this to you?!

Carlsen is not going to be *right **and*** his opponent the one most wrong person ever in history, the time the one time when he is wrong.

And neither are you going to be right that your opponent is the most upsetting wrong person ever, objectively.

And you weren't the World Champion in the first place!

2.6 Going round in circles

One cannot answer back against objective truth that "can we not do what we *think* we know", and say that is 100% certain it's 'the' answer objectively because 'the answer of your answer' is wrong 50% of the time and we possibly live in a mathematical paradigm where somewhere precisely in the region of 50% to 49.9% is possible, factoring in the proximity of the fewer errors when you are 'more relatively intelligent' are more consequential against stronger opponents, which a 'less relatively intelligent' person would never have demanded the right to make, since they didn't think they were a match for the higher levels to best in the world. The answer can only be **the** answer not what you think you know; that's **your** answer.

It may be a game we can't win. Nobody can achieve 50% or over positive outcome, factoring in consequence. An "unlucky numbers game" at the roulette wheel of the casino owned by number itself.

One cannot also say "doesn't my (relative) ability count for anything?" since you are not going to be *right* when you are wrong. And we will think we are right *the most* when we are wrong.

So it is going round in circles against the objective truth which must always remain objective, and uninfluenced by personal conviction of one's own thoughts, to argue either of the following:

- "everyone else thinks it" about what you think, on the individual issue. 99.9999% of people may well do. It doesn't make a difference. It wouldn't even if it were 100%!

Or

- "everybody else does it" about the *general* principle of acting on whatever they think, (on any given individual issue). They actually don't!

Most people, some 96% statistically, don't actually do it because they don't have clinical sociopathic tendencies, and for example more practically almost all people won't think they're close enough to the world number 1 at something, or important enough to risk projecting against the world number 1 *(whether they would do if they were* they close enough or not). So you would have to think a lot of yourself, or have psychopathic disorder, or preferably both to think that your own answer is **the** answer that **decides**.

Only people who think they're "important" and/or have psychopathic disorders actually insist it's their own answer that decides 'objectively' the answer, not analysis correlated to what the external answer may be (and what it is, when we get it)

You have to mitigate against external consequence outside your own subjective answer.

Nobody knows what they don't know.

1. Self-righteousness of insisting you know "the" answer because it's your own answer, is not an intellectual argument.

2. Nor is it an excuse for being wrong afterwards. "I didn't know" meaning that it was your own answer.

Putting those two together creates a matrix of mad Roman emperors. Before the external answer "I know the answer!" yet after the answer was against you "I didn't **know**-o-o-o-ow (different stress) it was the answer!"

You must take accountability for your own (potential) ego.

2.7 Social Prejudice Ego

A lady I know who is a so-called "Lady" in the House of Lords, has a different, somewhat less algebraic (or even less theoretical) approach.

The Lady thinks that "intelligence is 100% genetically based" and that she is more intelligent innately than the likes of me, for example. Prejudicing the only variable that exists from the start, before the start. So far so good!

I would consider her theory that of the sociopath, and she the sociopath's sociopath.

It would also fail to take accountability intellectually for being wrong in the past as they're "too important" and were doing what ***they*** thought was right (and they're the most important etc.).

They fail to mitigate intellectually in the past for being wrong in the future as it would be impossible as they're "too important".

The same mistake now, as then.

This is to say the equivalent of "everyone is like that, in thinking other people are the most wrong in the past, the one time when they're the odd one that's right so *we're* not accountable for it, and *I'm* not accountable for it".

Well the trick *is* to get it right though isn't it. To attempt to do that by analysis, and take accountability for one's own errors while, or preferably a little before, committing to external action with any negative consequence.

So

1) the person that's responsible for mitigating against one's own errors is oneself, not the rest of the world.

2) the person that's responsible for mitigating against somebody else's own errors is also *them*selves, not the rest of the world.

NOT

1) I'm not responsible

2) and (or because) "nobody else is responsible".

I wonder what might have happened to the Lady as a child so that the taking of responsibility, traditionally learned from parents, seems to have been psychologically overridden by intellectual abuse perhaps suffered by her by those who should have been responsible. Actually a large number of people in such positions if you get close to them say "nobody ever stopped other people doing those things to *me* when *I* was young" and "it's *my* turn *now*!"" wonderfully demonstrating the psychological problem of displacement.

And whether you *do* happen to get it right subjectively or not is not so much the point, as that only 96% of people know that they have to possess the intellectual self-awareness that one *might* not. The other 4% of people think they have no further responsibility of gatekeeping but to decide their *own* answer of what they are thinking at the time.

So the criticism of the Lady's theory is not that she makes errors, and I want an impossibly clean run of the horse in the show jumping every time, but that she fails to acknowledge that she can or will make errors, so we get leveraged risk, the most leveraged against the worst people to do it to, in the worst way, at the worst time!

1. She needs to understand that your opponent can be right, and mitigate against it.

2. Not to not understand that your opponent can be right by a wrong understanding of thinking that your own subjective personal thinking decides. That *you* have to know, when you're as important as the Lady, because *you* are *you* (subjective relativism of being relatively intelligent) but your opponent, were they conceptually more relatively intelligent or right than you, doesn't have the same right simply from your own lack of objective perspective. Where in reality both of you are x conceptually, you are not A or White in the chess game and they not B or Black conceptually. For if they were, the Lady would be the first player who decides as White, but also the player to go first and decide as Black. This can't work in the conceptual or impersonal (truly objective) objective. It cannot be so! If anyone was the real person who should decide, it would be the other person when they're right, but assuming they're not making the mistake against you they won't try to dictate the answer. So the mistake is not even so much the fault, but the perception of our own faultlessness (as judged by ourselves, where we could never perceive our own subjective error, subjectively).

If you don't know something, you are not excused taking external action with unlimited consequence because you're "relatively intelligent". Relative intelligence is not relative intelligence against someone that is more relatively intelligent than yourself, and you have to mitigate against that possibility where you can only use non-subjective sources of information. You're not excused the non-mitigation. If one were, anyone with influence can put World Champion Kasparov into a fate worse than prison for 40 years if you are Nigel Short. And Short in that fate if you're Julian Hodgson (a candidate for World Champion). That's not objective.

So the mistake is not one of not knowing, but one of not understanding that you cannot know.

To use my one joke/witticism I can tell at dinner parties about/to finance professionals: it is like the old banking derivatives story from the investing industry.

"There are 2 types of people:

1. Those that *know* that they don't understand banking derivatives

And..

2. Those that *don't* know that they don't understand banking derivatives."

It's not a case of the other person simply was definitely trying to upset you (irritation). They could be more intelligent than you (intellectual annoyance).

To not understand that you cannot know is *self-righteousness or sociopathy.*

To ally it with social prejudice creates an almost comical void where self-awareness should be.

3. But our Lady is the s***ociopath's sociopath***, and her theory the sociopath's friend: by *actively espousing* her theory that:

a) intelligence is 100% genetic based

b) her family and similar are the superior ones in Britain/world innately.

Hence the Lady is always right with what *she* thinks and, for example, a mixed race person like myself would never be right against her because people like the Lady decide and especially the Lady, decides.

And now when it has gone wrong she's "too important" to be held accountable, and was "doing her best" in a world where she's never wrong in the past or present, and will never be wrong in the future.

2.8 Sociopaths

The sociopath doesn't see objectivity impersonally, but personally (i.e. subjectively) "as objectively as possible" what the more important person in the situation thinks. They think nobody else is as interesting or important as them, themselves, because they are them. so it will be *their* own answer that decides 99.25% of the time, and if King Charles is the other person they will go along with King Charles' decision, and bide their time to attack the rest of the population which they think they can justify they're more important than them to themselves. A sociopath may well think he gets his hands dirty where nobody else is prepared to, and has glorious fantasies of his own grandeur since if it is his own answer that decides, that can be a way away from reality.

TH in 2006, a British diplomat, a man with an opinion of himself that would put Icarus of legend to shame, clearly didn't understand his own 'famous' 'Richard O'Brien from *The Crystal Maze*' example at a party in Bolton.

He was explaining that he as someone extremely intelligent was always being messed around by people who wanted to say they were right against him who thus wanted to play games with him they wouldn't win but thought they would. H was thus like Richard O'Brien on a more difficult Crystal Maze who had to spend 95%-100% of his time dealing with people who were wrong against him in arguments, and less commonly was he wrong himself. I pointed out that "I'm Richard O'Brien too", my father joined in with "I think I'm Richard O'Brien on that basis too", then another associate "I feel like I might also be Richard O'Brien from *The Crystal Maze* on the basis being discussed" like Spartacus. My associate said "you have to say 'everybody' to him (Mr H)" to try to get him to understand the point it happens to everybody relatively intelligent and welcome to also everybody else's world that's intelligent as well as your own of what every single other intelligent person must put up with by natural law that other people don't know they're wrong against you subjectively. I told him it's everybody who is relatively intelligent suffers the same timewasting fate. After telling us he wasn't "everybody" multiple times, told me off for talking about myself when it was about him, then when everybody agreed with me said "then it's everybody then" and why did I think it applied to me personally. I pointed out that he mentioned himself, so I mentioned myself conceptually i.e. that happens to me, that happens to Dad, that happens to your father, that happens to Nigel Short the chess player, that happens to the chasers on *The Chase* with Bradley Walsh, and that happens to people from *Child Genius*. That I only mentioned myself conceptually, because he mentioned himself as if it only happened to him. Then I mentioned "everybody" if he didn't like the conceptual use of self, to explain it literally.

H was labouring under the delusion of the self-righteous or sociopath that nobody else has the same problem. On being pointed out to him he was totally bewildered as to what anybody else was talking about. "We'd like to now stay on this subject" said my associate, a psychiatrist, but they were not given time to prove H's sociopathy. And we never saw this British diplomat again. I think he's a senior MI6 Officer now.

His thinking is pure self-righteousness. Thinking he is:

1) attacked by people less intelligent than him, who deserve to be punished for their self-righteousness

2) has the right of being self-righteous to people relatively intelligent to himself, because he "didn't know" but wanted to insist he did because he's relatively intelligent (but not as relatively intelligent as the people he's wrong against) because he's *important*.

He has the right to do it to other people as he's subjectively relatively intelligent, but if anybody else did it to him he'd want to put them in prison or worse. So only he has the right to do it, nobody else! And it's the same thing! And that guy can't see it he's too wrapped up in himself.

I happened to be privileged to attend the same school as the former Director of GCHQ, Brian Tovey, though in my case from winning a scholarship from state school. Tovey who believed in God and was also somewhat well-described as "romantic" personality in the literary sense, and gave marriage a go on more than a couple of occasions, said on H, the same British diplomat in 1992: "I 've no doubt the young man is a sociopath and Lord knows we've had more than our fair share of those in the British Government and we all know the damage they can do".

H said "no need to bring *him* into it, Brian" (The Lord) and said "and if he can't stop me, **then maybe I'm better than him anyway!**"

2.9 Circular Logic

Having watched such a sociopath as H get things wrong that were objectively (outside his subjective knowledge) 100% absolutely theoretically certain he was wrong about 24 years ago then say 24 years later "I didn't know what I was up against" and "we didn't know what we were up against", even though he was told precisely 24 years ago "you personally don't know what you were up against" (the point being to mitigate) is a moment as comical as it is unfortunate.

In the original argument, he didn't mitigate against it because he thinks his mitigation is within his own personal (what everyone else calls "subjective") belief. 96% of people think objective is outside his or their own personal belief, and therefore that he didn't mitigate against it because he didn't know subjectively, is an *entirely* sociopathic argument that does nothing intellectually and is entirely circular objectively.

1. I insist **I know** the answer, because the answer is my own answer, in 2001.

2. I insist "**I didn't know**" the answer, in 2001, in 2025

The first "the answer" is *his own* answer

The second "the answer" is *the* answer.

He's moved the goalposts.

Claiming you know the answer, if you think the answer is *that* it is your own answer, because you know what your own answer is. Claiming 24 years later you have no responsibility for any consequences of insisting your own answer even against theoretical certainty was right even though *the* answer was the other answer. Even despite now the absolute certainty it's the other answer. He claimed Google could never be worth £1m (2000) or £200bn (2006) in the future, and he was 100% objectively certain it was "the" answer (but meaning his own answer that he thought it, since he has a psychopathic disorder according to multiple psychiatrists). Well.

It merely *highlights* the *clinical psychiatric* issue of Mr H that would require more than years, sometimes 1-2, probably 3-5 years in his case, of a world class team of psychiatrists to correct.

Other people knew Mr H was wrong to be 100% certain it was "**the** answer", and 100% certain the people behind Google were "the most upsetting wrong people ever" simply because it challenged his own answer. But Mr H insisted he was right, "absolutely" "in absolute terms" we checked what he meant.

Mr H claimed he knew the difference, when it was pointed out what he was saying was a clear psychopathic disorder, as he didn't decide the answer objectively. In hearing that he went down to "99.9% recurring percent then, that it's the answer, or as good as" but looked like a toad about to explode, his neck pulsing with barely concealed rage, and later flew off the handle at the psychiatrist for daring to challenge his own answer down to 99.9% recurring saying "I was 100% certain it was my own answer". "*The* answer, not *your* answer Mr H remember we agreed?"

That turned out to be a lie that he was estimating **the** answer. And the lie of a sociopath who doesn't understand the difference between "100% certain in my own mind" and "100% absolutely proven".

So the non-sociopaths speaking with the sociopath had to defer, because of non-sociopathy, to the sociopath who didn't know he was a sociopath.

But the sociopath can go round in circles and 'shift perspectives' like a shapeshifter.

"I know" I'm right against theoretical certainty, but "I didn't know" for even 0.0000001% blame for being wrong.

Only if you're a sociopath who thinks that "you're you" is the only point anyone ever looks at objectively, so that if you thought something at any time in the past it is "the answer".

From our objective perspective that is a shift. Why would his presence in speculation before the external answer make all the difference, but his presence after the external answer in certainty make no difference whatsoever?

I know that if they were reading this, they would both argue "because nobody can ever do anything else".

And, once again, it would depend whether you thought mitigation meant *inside* your own personal opinion or *outside* your own opinion.

I have made a few *Knightmare* (television program) analogies in *The Simulation Code Revealed* but another one is whether you think objective is:

1) a world where everybody does what they think under the helmet,

and 2) that is what you should do too.

Or:

1) a world where 96% of people mitigate against what they think under the helmet and correlate to the external world taking external advice from clues and guidance and theoretical certainty (and 4% of people have a problem)

and 2) that is therefore what you should be doing too! Not simply what you see yourself under the helmet, even though it's 100% certain outside your viewpoint you're obviously absolutely certain to be wrong!

If it means **inside** nobody need ever do anything but what they think under their own blinkers. And the external world should expect zero mitigation and zero responsibility outside those thoughts under the blinkers.

If it means **outside**, 50% of the whole *point* is to **mitigate** *against* your own opinion when it is, for example, 100% absolutely theoretically certain it's the other answer. You mitigate or "gatekeep". You are the one that has to accept responsibility for the heinous consequences of your own intellectual errors creating external consequences.

Quiz:

Q: Who is responsible for the external consequences of one's own external actions when it's you and you affect somebody else?

A1 applying to self: oneself?

A2 applying to self: "nobody" or "the rest of the world should have told me" since you didn't know (since you got it wrong)

Q: Who is responsible for the external consequences of one's own external actions when it's somebody else who is not you who affects you?

A1 applying to others: them

A2 applying to others: "nobody", they didn't know. "the rest of the world should have told them in a way they understood" so it's unfortunate you got put in the gulag for 30 years as the USSR's best chess player, but both the postman and the taxi driver didn't know. They were the best chess players in their village in the Communist Party. You upset their own answers at chess the most in their lives.

You have to **decide** which of A1 or A2 is your philosophy. You have to have both A1 or both A2, one way round or the other. What applies to the goose must also apply to the gander.

Hence going round in circles vs sociopaths. They don't gatekeep, because they believe their own answer under the blinkers of their own personal opinion is "the answer" (their answer and the answer are synonyms) and hence infallible, in terms of what we go forward with.

And uncriticisable after the event. "I did what I thought was right". But did they mitigate against the greater evidence outside their own personal opinion? No. (A2 applying to self)

Did they understand the *concept*, even, that they *had* to *do* that. No. They believe everyone else does that (A2 applying to others)

So the idea Google could be worth £1m (like Kasparov being right against the postman at chess) was the most upsetting wrong idea ever "objectively". Of course it was.

2.10 Non-Existence Ego

Accepting that in any intellectual argument that it is one person's ego or the other but could be either objectively: this is one of the key starting premises of this book. The premise is however not accepted by everyone. Where the premise is accepted, it is 'unlikely', as in completely impossible, to be the ego of the person that didn't exist at the time. For example a racial puritist named Mr H makes decisions against people "because I have to" insisting other people are wrong when they're not born or for being born. This claim made against the alleged wrong person, person B for "sullying racial purity" by being mixed race could only intellectually be the ego of person A making the claim. The decision had been taken before person B's own birth, thus could not be person B's ego.

What Objective is not

This book was written because 2 people, Lady Mystery-Brit and Mr H, described my speculation that Google might end up defeating Microsoft in search in May 2000 as "the most upsetting wrong idea ever objectively" and myself "the most upsetting wrong person ever objectively" and estimated both answers as "100% Google will never defeat Microsoft" and "100%" I'm the most upsetting wrong person ever "objectively" for saying that it might. That's not objective. 'Objective' correlated to their own personal view is subjective. I realise I have been nastier in this chapter than others, as it really is necessary to challenge such sociopathic thinking, and to have any effect it must be stated clearly.

These people like the Lady and H don't seem to understand that while it can be red or black at the casino..

Correlated to their perspective from red can never be black and from black never be red. So we have A1 A2, B1 and B2 as possibilities that have to be dealt with, not just seeing the possibility from the bag of white marshmallows A, or the bag of pink marshmallows B. Without stating the obvious too keenly, inside a bag of white marshmallows, the marshmallow is always white, so 1 pink marshmallow would always be an upsetting wrong answer *correlated* to it. Inside the bag of pink marshmallows, the marshmallow is always pink, so 1 white marshmallow would be an upsetting wrong answer correlated to that. Correlated to the insistence the ball will spin red at the roulette wheel, black is always an upsetting wrong answer, correlated to the insistence the ball will spin black at the roulette wheel, red is always an upsetting wrong answer. You cannot not be upset by it when you're wrong. It literally upset your answer.

Clearly nobody would ever be wrong if objective is your own answer as what they thought was right was always right. But as Mark Twain said "it's not what you don't know that gets you, it's what you think you know for sure that just ain't so".

So the point is not that you thought it likely 100% to be red and that that means something because you subjectively think it, and you are relatively intelligent. But that people who are more intelligent than you, of which there is always at least one person in anyone's lifetime because the world chess champion gets old eventually, will come across as the worst and least likely black to you if you're going to prejudice it. Where in fact he was metaphorically red and you were metaphorically black anyway, or vice versa. You are the thing you think your opponent is. It's your ego, projected onto the opponent. Whatever you think they are, you are 100% of it, and probably 200% of it at least for insisting it's the correct innocent person who is the incorrect gentleman that you are.

A failure to understand this rather simple A1, A2, B1, B2 set out of outcomes with 2 variables where the only variable variable as it were is not just your own mind being incorrect but a disorder. You have to appreciate that you *could* be incorrect and to manage your own mind. You have to see it as if from a neutral perspective, like a bookmaker, and conduct objective analysis like that, even or especially if subjectively involved. Not to just "act on impulse" of the impulse of your own thoughts themselves.

Dealing with people more intelligent than you when you don't know they are, is like discovering an alien artefact in a science fiction film where you discover a modally unfamiliar form of input such as you are being spoken to telepathically by aliens.

You don't trust it as a piece of input. It's strange. Because it logically follows you would never have experienced telepathy before.

This shouldn't stop you from understanding the thing you were being told telepathically by the aliens, and that it may therefore very well be correct. We can estimate the chance that aliens are friendly and telling you the truth at very approximately 25% to 50% let us say.

The refusal to even accept something unfamiliar to your own thoughts is understandable. Is it a possibility since we "don't know that"? It is much more socially easy (and acceptable) to 'throw your toys out of the pram' about having to deal with something as complex that breaks all the usual normal rules that worked almost 100% of the time in the past. Nobody else will challenge us on it. Nobody else "knew" it either.

We probably won't face aliens, but the point is to elaborate with an extreme example to exaggerate how someone who is a genius or highly intelligent may come across to us. I might have predicted that Google will defeat Microsoft in search at odds of 1 million to 1, but what about the people who came up with the idea for Google and were able to defeat Microsoft at odds of about 7.5 billion to 1. Whatever genius idea they may have, new novelties, none of the rest of us can see it. But to condemn them before the spin, since it upsets our own idea that Microsoft will defeat Google in search (and everybody else's idea but a few!) is not objective. That pools panel decision will end up intellectually crucifying the founders of the company 500% bigger than anything else in the UK in history.

As a side note, since most – or all of us – as far as I am aware have the impulse to do this against humans 0.1% more intelligent than us, God help us dealing with actual aliens who are more intelligent or in any way more than 0.1% different to us. I can't see it ending well.

*

It seems almost childish, like a temper tantrum, the Mr H approach of deliberately leveraging of the "I want you to show it to my subjective narrative, in a way *I* understand" "or it's simply the most upsetting wrong idea ever" is a demand that won't cook with the truth that the aliens possess.

To then do this the other way round, in a spiteful interjection of ego: "*unless* you prove it to my subjective narrative, I'm going to leverage *against* the *possibility* even of truth with *my* ego". That is truly what we are doing. Because the answer is red or black and thus there is no bailout. We either go with the ego or against it. To feel the ego, to feel these emotions is to understand theoretically that the other answer is theoretically correct.

The "stick in the mud" "I'm not budging unless *you* prove it to *me*" runs out at baseball everyone more intelligent than us if we have any power over them when they were going to hit an incredible home run. This form of positive intervention is everywhere in society. When you work for an organisation that has force, like the local council, if you've ever dealt with them or a staff member who wants to be difficult. This is all over society, and it's all over intellect. And almost 100% of your life would be spent guessing how unintelligent other people are if everyone was like it. But only 1% of people are psychopaths and probably another 20% or so are jobsworths. We all experience this in life practically, where we know those sort of people are out there, and we feel a mild sense of relief when the person that has authority over us doesn't exercise the authority directly like a dog with two dinners. We think "I'm glad they're ok". But then also we occasionally come across the ticket inspector on a train that doesn't accept you've left your ticket in your bag on your seat when you've gone to the toilet, and when they've asked you for your ticket, and you've said that, and they reply "but you haven't got your ticket on you", any attempt to explain further is met with a further statement of the technicality that you are supposed to have it on you and therefore "you are travelling ticketless because you are not in possession of a valid ticket".

As the American Libertarians say: staying out of someone else's business is the overwhelming number 1 thing we should all do for each other. You'd have to be God if you're going to intervene on other people's decisions for themselves and hope to get it right all the time, and I don't know that I fancy even his chances...

Later using the social excuse "how am **I** supposed to know that thing was possible that was 100% certain but the input was not to my liking" "anyone else would not have liked it". Well **you** intervened didn't you? We told you not to (I ***just*** said, God couldn't do it!) but you did!

The impossible to understand alien telepathy in real life is not telepathy, but someone else being right in a way you didn't understand subjectively. Which is every time someone else is right against one in an argument.

Thus they were red and you were black where the ball spun red. But 1 million times in 1 million other times you would have been right.

Like Glenn Roeder said West Ham would have stayed up "100 times in 100" in 2002/03, but he still relegated us.

Chapter 3 - How do we know it is our Ego?

3.1 Subjective vs Objective Relativism - the Mathematicians Example

3.2 Mirror Example

3.3 Coin Toss Example

3.4 Absolutism of Most Enlightened

3.5 Horse and Jockey

3.6 Zeebrugge Ferry Disaster Example

3.7 Flying a Kite

Anatoly Karpov the chess player highlights an example of the difference between objective and subjective relativism.

Karpov learnt how to play chess sideways on, as a child, first playing wargames and various other games on the chess board. When Karpov got round to playing chess he was not White (a) or Black (b) but Mr X, a birds eye or sideways on observer where he never identified as being White or Black. This is how he learnt the game. Karpov is considered the most objective player, in the most intelligent sport, and was World Champion.

That difference between:

"I am White" (A)

"I am Black" (B)

"White is White" (X (but A))

"Black is Black" (X (but B))

is important.

When Karpov is playing someone better, where being Black would represent having to equalise, he (since he is not as unobjectively "he" as most people, but more akin to an independent observer by perspective really, having literally trained himself to think like that as a boy) is prepared to do it, and to not be distracted with ego arrogance.

Anyone who thinks in subjective relativism, such as against an unknown they are probably better than, is going to come unstuck the moment or one time that person in better.

Even if you suspect someone is better than you, chess would teach you against trying to "control" it, which is another great point chess makes to our egos. We should not be really aggressive with Kasparov to "control" his world-class attacking aggressive play himself. We should also not sit back in a chanceless position against Karpov, the master of maximising small advantages. Both these approaches make it more about us, we through our ego are very "present", whereas the ebb and flow of neutrality would suggest the entirely different but uncomfortable approach is the better one. Our ego always wants to make its presence felt.

3.1 Subjective vs Objective Relativism - the Mathematicians Example

Once we understand the hypothesis that we are in some sort of simulation with rules (Chapter 1) we only need to know that *ego* is our guide of what *not* to do (Chapter 2). However, not so simple, as we now have to determine whether it could be someone else's ego or still ours (this, Chapter 3).

Yes other people have ego, but so do we.

Yes other people appear less intelligent than us, but so do we to some others.

We cannot use the idea of subjective objectivity (I am person a) where I am x, and you are x and another human being, say mathematician Hannah Fry, is x to insist that we are relatively more intelligent than someone else "on the balance of probability" where that probability is inversely correlated objectively against someone more intelligent than us.

Let me use the mathematical example of a game where 3 mathematicians stand triangularly apart from each other looking at the other 2, and each one of them has either a black king

or a white king from chess on their forehead. Where this is a puzzle that has an answer, and the answer is logical and not a trick, could we work out the answer to our own colour

by looking at the other 2?

Well if we approach it 'subjectively objectively', I look at the other 2 mathematicians and they have say 2 white kings. I can come up with "a myriad of linguistic reasons" or even "mathematical reasons in the linguistic mode" such as the most obvious answer may be my king is white if theirs is, or perhaps since there are 2 white kings out that there is a dearth of kings and my own king is black?

But we would never make progress rationally until we start looking 'objectively objectively' and look at it as:

- We are x

- The person we are arguing with, Newton, is x

- The independent observer Hannah Fry is x

Instead of

- *I* am a (looking out the pillar box slot of my own self)

- Newton/Hannah Fry are b and c

" *I'm* more likely on the balance of subjective probability to be more intelligent than Hannah Fry, (and by the way, I don't care what Newton thinks). I'm more likely to be intelligent

than some guy who is silly enough to get hit on the head by an apple.

Well the point is serious. Some of those people may have more right to be more subjectively confident than me, but neither me nor my opponent has the right to claim it with objective certainty on that subjective basis, and nor does someone else not involved, or either of the involved with each other with that bystander 3rd party (b and c).

If I start doing it, and it's flawed in my own personal mind, as it surely is against Newton who though I was relatively good at mathematics I am not relatively as good as the person

that's relatively better than me..

why on *earth* should *he* not be able to make the claim against *me but me the claim against him*? It's *baffling*. The lack of objectivity to "on the balance of subjective probability" claims. It's 'relativistic bullshit' that has no place in serious absolutist discussion.

This is *murderous* when conjunct to *ego*.

"I am *probably* more intelligent than Garry Weinstein some chess kid who is 6 from Baku" (World Champion Garry Kasparov). It doesn't work. It's nonsense.

This is the layman's terms what I meant by "inversely correlated". The person you're *wrong* against is *the most* annoying to *you*. So in *your wrong* mind - he's "the most annoying of all *wrong* answers", because you went on the "balance of probability" where that was merely *subjective*ly objective. It *was* an attempt to be objective as if only your viewpoint would ever exist, but it wasn't a *very good* attempt to be objective since you never actually considered it neutrally, as if the answer of the other person's vantage point might have some chance of being correct when you don't think it.

Relativism is a very, very bad philosophy.

It prejudices the answer from the start. The only thing that is actually a variable that's undeniably a variable.

Example of bad train of thought:

0. I think in relativism which is subjective objectivity and I'm very good at chess because I'm no1 in Wales or no73 in England.

1. I meet Garry Weinstein from the Soviet Union who is a chess kid. "I am probably better than him".

2. Garry Weinstein plays me at chess. I start to lose. Kid is "most annoying of all people I am better than and going to beat".

3. Kid does not quit! And thinks he's going to beat me at chess! I stop the game and declare kid the "most obnoxious of all people who are ever worse than me at chess".

4. Because it's *his* ego, he deserves punishment.

5. But because his ego was "*worst* of all wrong people - his punishment should be 40+ years in chess jail or similar".

6. It turned out 20 years later he both invented a time machine, teleportation and telekinesis devices in the Gulag and used them to become World Champion. "How was *I* to know, it's not my fault I didn't *know* he would do that!"

To summarise

1. If only our one chess/mathematics person existed, no Garry Kasparov, or Hannah Fry, or Newton existed, the loser that thinks they always win would be person x in universe y, where x is electromagnetic and part of y that is electromagnetic and produces ego velocity reaction z. And since nobody else exists, it must be the loser's/one's own ego to be annoyed.

2. If every other chess player (Kasparov, Karpov) or mathematician (Newton, Fry) is brought back to the world, but without emotions, then every time you are annoyed it is still your own ego.

3. Only since other people exist, and have emotions, can it be someone else's ego, but we must still remember it can be our own. Therefore it's worth bearing in mind that someone can still be more intelligent or better than you, when you get to this point. This is the starting point for calculation, not the end.

In chess, Russia sometimes reminds its players before the game that the opponent *may*, not necessarily is, but *may* be better than them.

I feel like some governments could do with this "reminder".

3.2 Mirror example

You are not the World Chess Champion. But if you were, you would probably recognise that there is only 1 board that decides who is the World Chess Champion and that is the top board.

All other chess players are only "relatively" good at chess.

And the person who is "the best, relatively" without being "the best" is

interestingly the worst placed by subjective perspective of anyone. Like the

racing driver in 2nd place in the slipstream.

Let's say board 1 is in Harvey Nichols London changing room but every other board is taking place in the mirror, metaphorically, of the 999 thousand million

billion reflections of the absolute board 1.

The absolute board 1 thus has reality, but every other board has ***proximity*** from the '*objective* objective' stance of White on board 1 that wins. They are in no way "relatively" intelligent

in any positively-related fashion.

This therefore "flips" where White on board 2, is Black on board 1, of your own subjective relativism. So the only person who is "the most unlikely, by a long way" from your vantage point in the mirror is actually *non*-coincidentally, the only one better than you.

This is why relativism is so unreliable.

And why anyone that thinks they're intelligent "in relative terms" is a goner, theoretically. It's almost the ultimate oxymoron: "relatively intelligent" because it means by definition

that there is at least one person more intelligent than you, who you will think the worst that anyone thinks of anyone if you are number 2 and they are number 1.

3.3 Coin Toss Example

Let's say you got a game for the Bolton B team at chess, but your captain Julian was the unluckiest tosser of all time in the coin flip before a match, playing a chap, Andy from Stockport who was the biggest most successful tosser.

Julian seemed to always call heads where it is tails and tails where it is heads. And Andy seems to always go heads when it's heads and tails when it's tails.

The reality of this outcome is still 50%/50% whether they face each other with their history, or Andy wins a trillion tosses against Julian in a row.

The trillion coin flip victories in a row by Andy, are just *proximity* essentially. The best trick the devil ever pulled, to make you think you're good at coin flipping because you demonstrate great practical relative ability by result.

But we do not invest our ego in coin flipping. We invest our ego in the thing that we define as **our own intelligence**.

So being "relatively" intelligent, is the equivalent of winning 100 coin tosses, or 1000, or 10,000, or 1 million, or 1 trillion depending on how much you have invested intellectually...

Which the devil will give you...

To set you up.

To then read anything, to take anything *from* this into the coin toss, anyone against anyone. Even if Andy has the best record in the flip in the world, when you play Kasparov or go to the top table in Harvey Nichols, it actually doesn't mean anything at all.

It gets *inversely correlated* against someone who is better than you.

This may not be a coincidence, if we are in a mathematically alive universe.

From the other direction; you may ask me, "I thought there was 50%/50% chance?" In coin tossing yes. But in relative intelligence no, since you and a myriad of other people ("everyone") have the intellectual investment in agreeing the "two chamber approach" of looking at your own answer and your own successes. Everyone else looks at their own successes too. Thus everyone else including you, relatively or super relatively intelligent, can spot everyone else's relative superiority and inferiority up to just short of the very top level.

At the very top level, what everybody else thinks about what works against everybody else lower than your level falls completely apart against the world elite. A cultural example that comes to my mind is the swordfighting in a PC computer game Monkey Island from approximately 1990. Against all of the standard pirates, there are a group of standard insults that will work 100% of the time, it's merely a case of putting the right answer to the right question relatively. Finally however you have to face "the Swordmistress" and absolutely all of the regular insults fail to work and lose, and you have to learn and create your own unique new answers to the problems asked at this level.

I often feel this is why none of the football teams I ever follow such as Birmingham City and West Ham ever have any chance of victory against Man City or at Old Trafford. They will play by rote trying not to embarrass themselves, and do what they always do. But doing what they always do, makes the defeat almost inevitable or certain. So Steve Bruce fails to win in 53 attempts, and David Moyes in 37 against big teams away from home, but playing the systematic relativistic approach quite often yields a 0-2 loss or 0-3 loss, rather than getting thumped 2-8 or 1-6 the way *trying* to win the game uncomfortably would do. Only teams that risk ego humiliation such as Ian Holloway's Blackpool, that give themselves 5% or 10% chance of actually winning the game at least, but much more chance of a heavy defeat too, ever come away with victory.

Thus your chance against someone better than you is less than Steve Bruce's chance of victory at Old Trafford, but you think you are Mourinho's more dynamic, intelligent, winning younger brother.

It is your own insistence of your own intelligence, born of relativism that kills you.

Thus you go into it with a confident 100% or 99.9%+ feeling of your chance of winning as Andy against Julian. But the "biggest Julian in the world" if you excuse the expression, ends up surprising you, as the inverse correlation also inversely worked against *you*, but at the *worst possible* time for you.

So to draw a graph: the graph goes swimmingly well and smoothly, before an untimely and horrific blip right at the end or just before it which causes a demise indescribable, as you thought you were the most right in the most certain of all right coin flips. Then lose. Oops.

This *approach* error, includes prejudicing the only variable (the variable being: who actually *did win* the coin toss in the **future**)...

Manages to create *three* errors when wrong.

1. You say Julian number 1 trillion and 1 who is "the worst flipper of all time" can't win when he actually does.

2. He is "least likely" against me.

3. He "upsets me the most" , by having: the worst attitude of all people who are 1) wrong and 2) arrogant against me".

The thing is *inversely correlated* down to *subjective relativism*. So the graph goes less and less intelligent equals more and more irritating apart from the *very most* intelligent, the few 0.00001% better than you that *spikes up* in *personally inversely correlated* fashion.

So if you look at a coin toss:

Whether you accept:

1: 50% chance heads 50% tails (absolute mathematical)

2: 99.9% recurring chance heads (i'm right), or 999 trillion coin toss losses in row by Julian at chess vs 0.01% recurring tails (opponent is right)..

It is still an uncertain answer in absolute terms.

And the 0.001% most impossible answer still contains everyone relatively better than us, <u>whether we admit it or not.</u>

You may not *admit* it subjectively, but it definitely still does objectively. So the 0.001% chance turns out to be a good deal closer to not only 50%, but actually 100% it will go wrong against you, when you deny it is possible.

So to deny it, and to deny that denying it is prejudice velocity, is exactly as circular and mad as it sounds, and **velocity**.

Then trying to answer that back by saying "but you don't know that we couldn't be *right* so it is *not* 50% or 100%" is excuse me for saying...

Not right mathematically. Because you do know the answer theoretically, because if it were wrong rather than denying it with velocity, say it was a million times more wrong than any other wrong answer and refuse to play the coin toss...

you would just toss the coin and prove it.

Instead we sent poor "Julian number 1 trillion and 1" the only coin flipper better than "lucky Andy" to a fate worse than prison for 40 years by intellectual prejudice, prejudicing the only variable.

Plus it's *you* that's insisting and not unlucky Julian number 1 trillion and 1.

This is a failure to understand subjectivity in relativism, even in being explained, still answer back to say that you can:

1) get away with insisting "lucky Andy" would win.

2) but not be held accountable for insisting that a "luckier Julian" *could* win.

In a 2 horse race/chess match that's inversely correlated.

A coin toss, heads or tails. With 50% chance each mathematically.

This demonstrates a lack of understanding (and listening skills) based on subjective relativism, even sometimes in the explanation of it being ongoing!! "*I* didn't know".

3.4 Absolutism of Most Enlightened

Similar to the cookie-baking woman in film The Matrix, the very most intelligent people in the world are never aggressive, and must think in absolutism. They have no backstop.

Russian chess players tell me that the very best chess players in the world are always defensive in nature like Petrosian, and attacking dynamic Kasparov was just for show! This bold claim possibly stands to reason algebraically, you can see you cannot be aggressive against the idea the numbers themselves always balance and are only *interpreted* into a secondary mode which does it.

So it's just everybody but the very best that are aggressive against the numbers (interpreting it).

Perhaps that's what ego is. The desire to do that. What a horrific error if it is to be "intellectually crucifying" the very most intelligent of all by insisting they're the only

people projecting themselves, but also worse than everyone else is by a million miles! Perhaps it's a singularity doing just that, and I discussed that earlier (singularity vs multiplicity).

3.5 Horse and Jockey

As soon as we are accepting the easy first correlation that it is one person's ego who is wrong in inverse correlation to the intellect of the other person who is right, we can look at the horse and jockey theory.

There are 2 horses, 2 jockeys, but 4 potential ways somebody can get upset. Jockey A can claim jockey B upset him by irritation, jockey A can get upset by jockey B by intellectual annoyance, jockey B can claim jockey A upset him by irritation, jockey B can get upset by jockey A upsetting him by intellectual annoyance.

There are 2 claims possible. Jockey A can claim it was jockey B's ego (and hence irritated him). Jockey B can claim it was jockey A's ego (and hence irritated *him*). But they might not be correct. And they certainly don't *decide*! (objectively)

Whoever made the claim, and whatever the claim is (he upset me by irritating me, must be the claim), down to the inverse correlation where only intellectual annoyance can upset our own ego, makes 4 outcomes related to ego from the 2 results of the race:

Horse A wins the race, but jockey B never claimed anything aggressively.

Horse A wins the race: but jockey B **insisted** when he got upset that horse B was going to win **and that's why he was getting upset**. It was *actually* jockey B's ego. jockey B was being self-righteous/ego-opinionated as there was a false claim by him.

Horse B wins the race, but jockey A never claimed anything aggressively.

Horse B wins the race: but jockey A **insisted** when he got upset that horse A was going to win **and that's why *he* was getting upset**. It was actually jockey A's ego. jockey A was being self-righteous/ego-opinionated if there's a claim by him.

An argument can't be caused by 2 completely innocent parties. If one party gets upset then he could be correct that the other person is irritating him. However, enlightened people would never make that claim **whether they got upset or not**, since the irritation caused by someone else's ego is best served by ignoring that person rather than hunting them down to get into an argument over it. Typically then, the wrong side in an argument starts it. Kasparov rarely beats a path to the door of a mediocre chess player, or phones them up to argue with them that he's better than them at chess.

It's inversely correlated objectively as easy as that. But your subjective mind 1) dislikes the people the most that are right and 2) compares them to the 'everyone else' that are correctly (As opposed to in this case, rarely, incorrectly) analysed as wrong against you. And you prejudiced the only variable that exists in the first place by insisting the horse with the blue rosette was going to win. "we thought it would" is no answer, unless the thing is *absolutely* certain. i.e. in *absolute* terms. Not "100% we thought so in our own mind" (above 50%).

Losing jockeys *always* claim that. You can see that everyone else in the world can claim it just like ourselves. But we only subjectively think it's important what the wrong person thought when the wrong person was us. We wouldn't accept it as important if someone with an IQ below 100 thought something wrongly against us, because they are them subjectively/personally. We wouldn't think it important if 100 people, or 1000 people, or 1 million or 7 billion people less intelligent than us also got something wrong against us, that they thought it subjectively before the race, and insisted it by putting us in prison for example. We would not find that a very compelling point of interest that they're them and they personally thought it. Somebody wrong can always claim it that before the start they thought it was the other answer and "just got it the wrong way round". It's what everyone thinks. How would anybody not think it? And everybody personally thinks the winning opponent against them personally is worst of all wrong opponents.

Our subjective opinion: meaningless objectively. Objective evidence is stronger.

3.6 Zeebrugge Ferry Disaster Example

The Zeebrugge Ferry example is another unfortunate example of from a different perspective the recklessness of 'subjective relativism'.

People are normally saying "we can't be expected to know that the 1 in a million chance of somebody being better than us existed seeing as how it's so unlikely on the balance of probability".

The people in this disaster thought it couldn't be reasonably expected that all of the other people involved in safely would also fail to do their jobs at 1 in a million chance, as well as ourselves individually. And then didn't do their job. On the *almost* certainty that the other person wouldn't *be* the thing that would cause a problem (being asleep in the ferry).

This is like a reverse version "I can't believe someone could be so stupid" so am not going to accept it as possibility (even though we were ourselves) instead of "I can't believe someone *could* be so intelligent" even though we ourselves are and appear so impossibly to people less intelligent than us.

And are now complaining it's the other persons fault for being the 1 thing "it *could* have been" to have caused the horrific disaster.

Someone can be so stupid that you can't logically predict the level of stupidity to not do their job in safety. The person more intelligent than you is not not doing their job by being more intelligent than you unbeknown to you either.

You should have some perspective on this especially because the only reason you're prejudicing the claim in advance in the first place, is that *you're* more intelligent than *other people* who are *less intelligent than you!* So it must be possible to be more intelligent than someone without them subjectively being aware of it. The objective objective, not the subjective objective that only applies the formula 1 way round to *yourself personally as the centre of the universe.*

3.7 Flying a Kite

Do you remember being young and attending American themed chain restaurant Pizza Hut or similar with your family? There were always paper placemats, with games on the back for the waiting children to fill in the time before the meal was served.

Often the game was a boy, a girl, and maybe a dog or 1 or 2 others have strings at the bottom of the placemat, and a kite flies at the top. *But by whom*?

The best, easiest way of solving such a tricky puzzle, is to go from the *kite, backwards*. Not to use trial and error of any order.

Where there is inverse correlation in every intellectual argument in adult life, you are like the boy the girl or the dog (3rd party bystander) where you will always think it is you that is flying the kite, metaphorically, but when it is the other party you think it is you *the most*.

Therefore you cannot prejudice this by starting at the bottom, or starting with yourself when you are not *absolutely* certain. Meaning in *absolute terms* not "I say I am absolutely certain in my own mind".

To prejudice the only starting variable, by insisting it's the boy, that "I'm THE MAN" on the basis that it's usually me and I like me, and I'm interested in me, seems a bit...daft.

It's not an intellectual point that you are you. And the girl and dog have as much right to claim it could be them as a (especially if it *is* them, which you can't objectively prove it isn't).

If you are claiming it's you, because *you* think it's you, why can't someone right against you (say, the girl) claim *they* think it's *them?* And if one of you are wrong (sometimes the boy, sometimes the girl) why could the *dog* not also claim it?

So in any *group* of arguments we can be sure of messing it up eventually, if we are going to prejudice the only variable. But we mess it up against the person who is the most right, who is the worst person to mess it up against.

It can't be an excuse to say "but I'm relatively good" against someone better than you. When you prejudiced the only variable.

Do not prejudice the only variable.

That is *not* the "absolute certainty" you are right, it is subjective relativism – certain in your own mind based on your own experiences subjectively. But applying it permanently leaves the 100% certainty you will be wrong theoretically and will be practically if you take enough risks by applying subjective relativism long enough.

Why go on about this difference? Because using objective relativism avoids the theoretical, and thus practical, certainty of the biggest mistakes in the world against the worst people to make the mistakes against (Chinese World number 1 at chess or Vladimir Putin).

Using subjective relativism, with 100% theoretical certainty, is thus only a matter of time practically before we make the world's worst mistakes, with the world's worst consequences against the world's worst people to make the mistakes against.

CHAPTER 4 - Difference between Upset, Irritate and Annoy

4.1 The Clock Example: the completely average person the pm is 50% and the am is 50%

4.2 Delineators

4.3 Archery Target and Board Example

4.4 Seesaw with 2 variables

4.5 Comedy Heckling Example, or X Factor, or Chess MATCH

4.6 Comedy Heckling gets TV program

4.7 Tweedle Dum vs Tweedle Dee

4.8 Red vs Black

4.9 Poison vs Antidote

When we have defined what ego is, and what relativism is in an *objective* context, we then come on to what is the definition that binds someone else's ego and one's own ego, and the phrase that covers both minds perceiving a slight.

I have to define this. But you can define it as whatever you like in words as long as you *stick* to it, and know you mean!

The only irrational variable to truth is *you* and your interpretation. When someone is more intelligent than you they are more intelligent than you. When someone is less intelligent than you they are less intelligent than you.

Like a chess player, you cannot go into an argument insisting that you are more intelligent than your opponent, like starting a game against an unknown opponent but insisting they can't be better than you generally or in the game.

There are 2 states of preparedness: prepared and unprepared. There are 2 results and hence 4 different branches. The one that is *murderous* is when you don't have a state of preparedness, but your opponent is better than you. This hits your *ego* with *friction* and "*intellectually annoys*". I.e. your opponent is intellectually (or 'chessically') correct. You are intellectually (chessically) *incorrect.*

Being incorrect but prepared would not intellectually annoy as you are not in a state of unpreparedness.

Once you get hit by your own ego in a state of unpreparedness your best chance at recovering is to acknowledge that a few seconds ago or a few moments ago when you lashed out that it was probably or definitely wrong if you know it was and to recover and regroup as best as you can.

If you leave it more than a few minutes to attempt the regrouping you may never recover to the correct position as it becomes that the other person is "the most 'annoying' never to be admitted by your mind again that it was actually your ego and instead a *drawn-out* process begins of going to the end of the earth or somewhere near it to prove the other person impossibly annoying.

People less intelligent than you and people more intelligent than you: both are upsetting, and both the less intelligent (could be you) and more intelligent (could be you) get upset in every argument. Every argument is caused whenever an *error* exists. *Both* get upset.

Introductory example: let's say off the top of my head in about 1990 the important future President of France Chirac wanted to admire a young boy's artwork, but the boy didn't want his portfolio to be admired, both are going to get upset from friction. And there is a question beyond "Chirac is important" which is there an unlikely (not seen very often) error present in our algebraic universe if Chirac got upset. Would Chirac usually get upset by this? Would Chirac usually get upset by someone who is wrong, more than a truth he didn't know making him wrong. So if he does, how is that even possible using the algebraic approach without it giving you the clear algebraic symbol/warning. Example of what it just told you is going to happen, the boy would have to become President of Russia, or the USA's leading strategist. Because there is no other algebraic/mathematical solution to that equation so the intellectual error was there, existing, on the part of the President of France.

So should we take this boy seriously, as a serious candidate? Since it is theoretically certain due to this algebraic theory, but not in the (usual) mode of language?

It may be worth doing this, or checking this, rather than have 8 billion people make 8 billion factorial errors, when we can at least check the mathematical way and find these anomalies that have a clear algebraic solution, and an absolute answer once we assume 1) the world is a simulation and 2) ego is a negative form of input. Even without taking it as gospel, we would still have the theoretical certainty. In fact we have the theoretical certainty *anyway* even in language as it's clearly an impossible anomaly, we all understand that. That someone can be more upsetting being wrong, to everyone, than anyone being right against them individually could ever be to each and every one of them. We are aware of the anomaly of the mathematical anomaly present. We just don't take it seriously because we never thought of not thinking solely in language..

If we truly do live in a mathematical simulation (where answers are played out in parallel universes by supercomputer calculation), it's absolutely telling you the answer (in the mode of mathematics) not just the theoretical certainty of it to be ignored in language. By the incredible anomaly itself, that was the mathematical information.

When Chirac and 1 billion boys with artwork exist, this intellectual error would still exist when (or because) both sides still get upset.

But it would be the other way round 1 billion times in 1 billion and one. I.e. the boy's ego, which would be irritating Chirac. But we **know** this time it isn't. That it can't be theoretically. Or mathematically. And this knowledge may have greater consequences if we start realising there is a *possibility* we must take it seriously because it seriously is a form of 'algebraic' answer in a mathematical world.

You can clearly work out this is the 1 time in a billion because it's the most upsetting wrong person ever in the minds of everyone, that has a velocity and variance higher than any correct answer. So it's completely obvious mathematically it's just 100% or close to it of humans think in language and 'go with' their own idea that the boy is the most upsetting of all wrong people ever, much more upsetting than any correct person, even though this is senseless algebraically/mathematically.

This algebraic way of looking at things (with the knowledge that we might be in a holographic simulation run on mathematics) is helpful. Infinitely more helpful than language, potentially, which will give you the *wrong* answer in ego whenever we are wrong, but mathematics will be able to convert that to give you the *right* answer with 100% mathematical and theoretical certainty by showing you how to interpret ego showing it is definitely the other way under those specific mathematical anomalies.

So you have to "think differently". Maybe there will be enough people, and/or a technological innovation in the future that enables more of the masses to do it. Perhaps this book is one of the first steps.

Until then we suffer the unfortunate "he's the most upsetting of all wrong people, more upsetting than any correct person" even though it's completely obvious mathematically it's the other answer. Because people don't think in mathematics, and won't allow themselves to be the one person to "think outside themselves" when "everybody else does it" (do whatever the hell they like, in language without paying attention to mathematical theory).

If these people had knowledge we were in a mathematical simulation (and there is great evidence and I will go through some of it in the final chapter) and they were going to get trapped here infinitely in a non-biological state folded into the simulation if they don't escape, they might be more vigorous in their desire to apply theoretical and mathematical certainty! Rather than say something completely impossible is certainly correct, and the think that what's certainly true theoretically is certainly the "most upsetting of all wrong ideas ever"!

Pay attention, you might be in a mathematical simulation and you would therefore not be able to *escape* the mathematical intellectual reality which already knows the answer (and already *told* you the answer) and you would have to take intellectual accountability for your own actions (or get trapped in 'the cube' permanently).

If you are struggling at this point to understand why velocity and variance demonstrates a clear theoretical and algebraic answer, just one we are inclined to ignore, it is difficult but not impossible to understand it, and I am going to give more examples in an entire chapter on its own dedicated to velocity and variance. A lot of people actually understand that it's theoretically certain according to theory, and we can all understand in layman's terms that someone could not possibly upset everyone the most in all of their lives by being the most wrong wrong person at something (say chess) and that be more upsetting to each and every person than thinking you're 3000 times better at chess than Nigel Short and then the actual talent upsetting you intellectually.

But if we live in a mathematical universe, not a linguistic one, where there is this infinite variance and infinite velocity and ego has to be an answer one way or the other, you know it's the other way because it could not possibly be the other person's ego. Whereas all the other billion times it could have been there, there was no variance and no velocity.

So the variance and velocity give you an algebraic answer.

IF

1) the world is a computer simulation

2) ego is the **communication** from the supercomputer, which knows the answer because it has played everything out (see 1)

You won't be able to refute that theory, and it appears to hold, which is maybe the strongest practical proof.

In fact if you know the theory, you can catch people out for fun if you know you know something they don't know, but they think they know. Then you will intellectually annoy them and this will upset them much more than irritation. But in their hapless linguistic state they will try to say you're the most irritating of wrong people more upsetting being irritating with your own ego than anyone who is correct and genuinely intellectually superior/annoying has ever upset them by intellectual annoyance which was *their* own ego.

*

Let me use an example of an am vs pm clock to show that the phrase "upset" - I am using - covers both am and pm.

But the phrase "annoy", more specifically "intellectually annoy" only covers pm

and the phrase "irritate" only covers am

and we start high noon, for every person equally.

4.1 The Clock Example: the completely average person the pm is 50% and the am is 50%

Since our first example is Mr. Completely Average, it's precisely equal the amount of times you are intellectually annoyed by relatively good or great minds is 50% (pm, and that's where we start with *our* ego) and the amount of times you are irritated by weak and frivolous minds is 50% in the am.

So the people *less* intelligent than you *irritated* you with their failure to understand things and *their* ego.

But the people *more* intelligent than you *intellectually annoyed* you with their hitting *your* ego that you don't see or understand.

This *intellectual annoyance* is like the hard pedal on a piano. It is *your own* ego and hits you *full square* starting in *your own* mind from *somebody else's* intellect.

The *irritation* is like the *soft* pedal on a piano. It is *somebody else's* ego but it still kind of triggers you because you don't want to deal with someone else's ego. But that is fairly *soft* by comparison to dealing with *one's own* ego which is murderous.

Imagine 2 castles with drawbridges up, one left, one right. If we fire something into the opposing castle of our own ego, hitting them, the effect we have by reflection could never be as much as our own ego. The false negative is never as much as the real negative.

So can someone be "so irritating, they become intellectually annoying"? No.

Someone can be "so irritating that we *feel* they are very *upsetting* where the very upsetting is like how upset you'd be if the milkman was jingly jangly with your orange juice bottles in the morning but a very extreme version of it. But he can never annoy you intellectually with it, delivering milk is not intellectually annoying to your own ego. So he can never murderously annoy you.

However if the milkman beat you at chess where you were England's no1 so you thought, but turn out to be number 2 to the milkman, whilst wearing an apple green t-shirt and you didn't know the milkman could be any good, then that would feel like the milkman trying to murder you. Or more accurately like *you* wanting to murder *him*. I'd rather complete my military service for the small European country I'm from (not telling) than suffer this murderous upset.

So you are comparing apples with oranges. Not oranges with oranges, or apples with apples.

So in summary: you cannot say someone is "so irritating they *become* intellectually annoying". They are *either* irritating (*their* ego) *or* intellectually annoying (*your* ego). But they *cannot* be both, and hence *cannot* be "so irritating they're annoying".

If it is only irritation, it could not upset you the most in your own life. It could not be murderous. Someone else's ego does not offend your own mind like your own ego being the most wrong it ever is, which is what ego is and what intellectual annoyance is and how everyone gets upset .. *the most.*

To say different would be to say that a "false negative" would be "more upsetting than every *real negative.*

I can make it even clearer by delineation and example involving X Factor, chess, Frankie Boyle and Simon Amstell.

I think the *real* issue is that people *forget,* and don't have the Russian Government to remind them before every encounter like the Russians remind their chess players before every battle, that your opponent can be better than you, and here's the real

issue:

- If you are in top 33.3% of the population instead of the top 50%, then you will not be right 66.6% of the time, you will be right circa 80% of the time. Because each time someone is less intelligent than you - of which there are now 2 for every 1 - then they don't just get 1 thing wrong 1 time. They might get 2 things wrong 2 times, or 1 thing wrong 3 times, or 3 things wrong 1 time, or 27 things wrong against you 182 times in total. The ratio is not a flat you are right 2 times in 3.

So you are not having relativistic arguments perfectly 50% of the time pm and am. You now have a situation where you are officially the dreaded "relatively intelligent". So your pm is down to 12 o'clock to before 5 o'clock, and your am takes that space.

If you are in the top 25% of the population: The ratio is now 3 to 1, of people less intelligent than you to more intelligent. So maybe in region of 90% of arguments you are right because you are right on multiple right things on multiple right subjects against multiple wrong people (compounding positively). And less people who are right multiple times against you. (compounding negatively). Those people who are wrong objectively, don't know they are wrong objectively *sub*jectively, hence you are accosted by multiple wrong people on multiple wrong subjects, possibly multiple occasions. Yes you will spend most of your life arguing with people who are less intelligent than you who want to insist they are correct against you (possibly more than once, possibly about more than one thing). Welcome to Richard O'Brien, the gatekeeper of the Crystal Maze's world!

How much can you expect to be in arguments to be arguing with people wrong against you who don't know they are wrong? If you are in the elite of our population:

top 20%: Ratio 4 to 1 Are we at 99% of the time factoring in compounding, maybe not yet.

top 10%: Ratio 9 to 1. I think we're at 99% now!

top 1%: Ratio 99 to 1. And now more so.

top 0.1% Ratio 999 to 1. And now MASSIVELY so.

What's my point? My point is *critical*. That once you get into the realm of the "extremely relatively intelligent" that we are all right "all of the time" virtually, or technically "*almost*" all of the time. And that's not a good thing for our humility in facing future arguments with people who might be correct against us.

So easy to forget, when the ratio of being wrong to right is maybe 1 million to 1 or 1 billion to 1 on the later numbers of being in top 0.1% of minds *relatively*.

But forget it you mustn't.

Do not forget *two* things together:

1. Someone else might be better than you.

2. If they *are*: *relativism* works *objectively* not *subjectively*. I.e. *they* have the same problem as you and *they* are entitled to it more, and *you cannot* claim "I am probably better on the balance of relative probability".

That is a *subjective* claim of relativism and not an *objective* claim on relativism.

You are casting yourself as A and your opponent B. And your opponent who could be better than you, and *is* in our example, could never be better than you in your mind.

You might be one in a million. He might be one in a billion.

You might be one in a billion. He one in a trillion.

You might be the world's best. Not forever!

No-one is entitled to do it.

To claim "I'm *probably* better" and *forget to play properly or at all the chess game*. Then claim later "I didn't *know he could be*" - the *same* error from the *other end (the end)*.

Start: I'm assuming I'm better subjectively: thus this guy is "most annoying" (when he's better).

End: I couldn't be expected to know that that one guy was better than me, because I'm so important subjectively relatively (also called the Scooby Doo villain approach – I would have been ok if not for that pesky kid that's better than me).

Anyway, to go back to the difference between irritate and intellectually annoy, if we can accept that someone, sometime, among all people, *can* and will be intellectually annoying us (they are better than us). Why then always assume they are irritating us (they are *less* intelligent than us) at all? Answer: If we don't know in absolute terms, we can't!!

It is not *that* difficult to work out, out of the moment itself, because intellectual annoyance feels *murderous* and would leave us having to say that the other person was somehow *inferior* to us.. *but at the same time* "the most *upsetting of anyone*" which would make an algebraic improbability so huge, even in layman's terms that the universe might explode.

You'd have to say it *linguistically* - as no self-respecting algebraic approach man would come *near* that claim – that the unknown kid at chess who upsets us even though we've already decided must be "the most wrong of all wrong things by a million miles" or some such.

Utter nonsense, and fairly easy to spot after the first 3 seconds. But to those who like to insist on the linguistic, non-algebraic approach, and claim not to be able to spot it, and that "why *couldn't* it be that"... here come the delineators!

4.2 Delineators

There are 3 delineators I can see if this is confusing. And no doubt it's confusing to people who don't take accountability because psychopathic governments have been doing it for centuries.

1. First delineator - the person has to be a professional level threat if they professionally annoy you.

So when you are saying "this person is the most wrong of all wrong people", how can they professionally annoy you if they are the worst of all lame amateur people?

So to be clear:

1. If you are insisting someone is worse than you

2. And they're upsetting you the most in your life/right up there

3. Which means of all candidates against you professionally they are clearly one of the worst candidates you've ever encountered and "the most wrong of all wrong candidates".

That is algebraically *your own* ego, since someone who was bad professionally like that couldn't pose a professional threat to you (oxymoron).

So a non-professional level threat, that professionally threatens every professional chess player? To the point where the contemplation of having to play the game is worse than any real loss/defeat ever suffered in any other game?

2. Second delineator - is it aggravating to discuss?

You say you don't know the difference between someone else's ego and your own ego being to blame for somebody else upsetting you?

But let's say you were heckled as a comedian, is probably one of the easiest examples as it's almost exactly a game of deliberately upsetting someone else in a sense.

If someone in the crowd heckling you upsets *you* and you are Frankie Boyle, someone paid to upset the heckler...

since the 2 people are inversely correlated in their fortunes - the heckler wins you are no good, you win the heckler is no good...

And you are paid to be there as a professional...

I find it a hard argument to take the idea of "the heckler has upset me the most in my life even when I could have them ejected and I will be going home after this performance in which I'm paid to stand here".. "but it's not my ego".

That it was allegedly the heckler is so bad at upsetting you, and thinks he will upset you when you know that he will not upset you (even though he did) and that's not velocity (even though it is) because "you're not wrong" (even though you *are*).

It's because the heckler's *approach* was so upsetting because he's so *lame* at comedy and so *arrogant* in *thinking* he would upset you..when he *wouldn't*... that *that itself has* upset you... and *the most in your life?*

Is that then aggravating to discuss in Gail's Bakery in South Kensington next week with Stewart Lee?

Is it: "I encountered this heckler, he was a silly billy" and laugh about it.

Or "I encountered this heckler, it is *still actively annoying me... someone's* ego and it's *definitely his* with complete active ego annoyance. But it's his. And he's gone back to his job without thinking about you.

Clearly when you have option 2, but say it's the other person's ego, this is not correct. How has the *other person's* ego actively continued to trigger the switch of your own ego when you are earning £100,000, live in South Kensington, doing your job, eating muffins and gone home 1 week earlier or 1 month earlier, or 30 years ago. Some people who were upset by their fathers in childhood, perhaps find ways for someone else to be irritating them 'more than their own ego and that still being active months later in an impossible way' in an act of transference.

Someone else's ego does not actively continue to aggravate you, weeks or months after any real threat to you from it has gone. Only your own ego refusing to accept it's your own ego and not the other person's ego. That's called "intellectual annoyance".

Your ego demanded aggrandisement, which didn't come because the other person was (perhaps rarely) more intelligent than you. This hit the trigger switch of your ego, and you didn't back down 3 seconds later or 3 minutes later, which was your best chance and easiest chance by a long way, like famous England chess player Danny Gormally punching world no2 Levon

Aronian then apologising to the Armenian team later the same evening. It becomes more difficult to do if it gets entrenched.

3. Third delineator - the Aliens example.

Do you attach yourself to the person for 30 years afterwards shouting "there are ALIENS out there" while attaching yourself to them? Real or metaphorical.

If you do. They are probably not some normal lame guy that's the alien irritating you if *you* are going to all of that trouble to attach yourself to *them*

You are probably the alien.

4.3 Archery Target and Board example

Algebraic approach reality is a refusal to *aggressively* interpret theoretical fact which is tautologous. To only behave with assertion *defensively* to *others* who are

interpreting it *against* us.

For example I do not claim that 100% White people are inferior to myself because I am a mixed race person as it is unproven. But if a White person insists on claiming they are superior to myself I would be forced to defend myself against *their aggression*. Whether I have any particular interest in arguing about it is taken out of my hands by an aggressive claim.

Most arguments are left/right on the archery target.

But to argue that algebra is wrong, and the opposite answer in a 2 horse race is correct and the horse representing the algebraic answer's jockey "most arrogant pro rata" because he upsets you the most with an answer you don't like, is not a left/right argument trying to find truth of the centre in the wind or flight can't be judged, but one that denies the archery board target even exists or that we have to pay any attention to tautology. i.e. "why *can't* it be this completely nonsensical answer since there are a million linguistic plethora of ideas"? Because that is subjective, and not objective. People who are being objective, do not insist their own personal answer is correct simply on the basis of what the thoughts floating fleetingly through their head in the moment are.

It is unlikely that a very wrong wrong answer, such as a poor archer firing arrows towards the very edge of the board with the least ability to hit the centre, would upset your own brilliant archery prowess of hitting the centre, throwing you off the most in your life.

4.4 Seesaw with 2 variables

When the linguistic thinker is saying that algebraic answer is incorrect there are 3 variables.

1. They are prejudicing the only variable at the start (is it A or B, I'm saying it's B thus it's B)

2. Then A upsets me the most in my life.

3. But A it is also, I insist, the weakest of all wrong answers against B the definite truth.

The girl on the left of the seesaw and the boy on the right. It is prejudged that the boy is heavier without absolute evidence, in something that upsets your mind to consider intellectually that your precious girl of an intellectual opinion could be heavier (wrong).

Then the level of upset represents how far back it is.

And how likely the answer is in your opinion represents the velocity (weight) from that position.

You can see that the answer doesn't balance, where every other correct answer balances.

But it *would* balance, if you accept the girl is heavier.

Which is the one thing you wouldn't accept at/before the start. Your idea could have been wrong (it wasn't proven absolutely).

Albert Einstein was adamant that when your own facts don't match the theory, you should change your self-declared facts to meet the theory, not change the theory to match your unproven 'facts'.

4.5 Comedy Heckling

If Frankie Boyle gets "upset the most in his life" by a heckler at Boyle's own comedy gig... it could be Boyle's ego or the heckler's ego, but it is unlikely or impossible to

be the heckler's ego.

The 2 are engaged in a battle to upset the other. But it's Boyle's profession.

Thus in the 1 in a million chance Boyle gets annoyed by a heckler professionally, rather than the other way round (the professional putting the amateur in his place which is what should happen) then perhaps the heckler is good at comedy? Perhaps a future professional if a young person.

It's either Boyle's ego.

Or if the heckler's ego: the heckler must be "the most talentless of all hecklers (1), and think the most of themselves (2) or some combination of the two (3) to upset Boyle with the *heckler's* own ego, the same as our see-saw example it doesn't really hold.

Most likely linguistically on a layman's terms basis, and in algebraic mode certainty, is that it's Boyle's ego, though he may be the best comedian in the country apart from this imaginary heckler.

It wouldn't happen in real life... well it wouldn't to Boyle, but it did happen (so some people tell me) to the CIA in real life under the leadership of John Deutch, where the CIA Officers were getting upset the most in all of their lives by a boy from a European country and said it was the boy's ego in that ridiculous *approach error insistence* of "we're right without proving it" and he must be "the worst of all wrong people" or "think the most of himself" or "some combination of the above".

See Case Studies (Chapter 8) to see if it really happened.

Sometimes it helps to see things from 2 perspectives, like language and algebra.

You could leave poor Frankie Boyle alone with imaginary hecklers, and say the same thing about the singers in the X Factor auditions as amateur comedians.

The combination of their lameness and arrogance, makes them "the most upsetting in my life" to professionally correlated Pavarotti?

Truly? Really? That's quite funny to think it might be true, we are all smiling at the idea. Yet we'd take it seriously most probably if we were the professional. Obviously if it happens to someone else, not us, we'd find the statement ridiculous. Of course the most talentless arrogant people on the X Factor are not the most upsetting wrong people ever that affects the lives of Adele or The Beautiful South. I find them funny myself. I try to imagine someone worse than me at chess, singing or comedy upsetting me the most in my life for minutes, hours, days, weeks, months, years and decades after, when I can simply go home and count my millions if I'm that good, most of us would probably be alright with the millions of pounds and global fame and find the idea it upset us the most in our lives unserious as a proposition.

But then I didn't prejudice the only variable by insisting I am better than everyone else individually in the first place who I think I'm better than. A prejudice that is inevitably coming unstruck at some stage, and usually only employed by people who are closer to the top of those professions than me, though you witness more overt delusion in action in X Factor audition first rounds.

It seems to be only people who *are* making that insistence – which is a failure of thinking in the *subjective* relative and not objectively and applying the same standards to others.

Either *both* of you have the right to claim you're better, and the person better than you has the right to claim it.

Or *neither* of you.

To claim it's only *you* because *you're* good – *subjective* relativism obviously eventually comes unstuck, but a theory espoused by Aleister Crowley, and followed by the likes of modern Luciferians in their lodges, seems to propose it!

Crowley wrote "what thou wilt is to be the whole of the law". One of his disciples said to me that thoughts create reality so all anyone and all of us ever have to do together is to work out what we think, act it out, and if we're all wrong we will learn something. Such as Russia don't like being called "the most upsetting wrong people ever" and respond with 3 nuclear threats specific to Britain, for example.

This disciple is a real advocate for such insane thinking and says that someone else who upsets her when she's prejudiced the only variable not only *could* be most irritating of all people, a false negative stronger than any real negative that's ever happened to her (or anyone else) but *is* because *she thinks it* personally. That just the thought of someone unfunny or bad at singing, though we don't know absolutely because she prejudiced it, is worse than any negative that's ever befallen her. And the only "unreal problem" that is worse than any "**real** problem", in human history. This clearly could only ever apply to *her* (subjective relativism), and not impersonally as could be demonstrably observed by anyone else in the world (objective). This is how 4% of people think. Psychiatry would declare it a clinical psychopathic disorder, nobody is denying that. Not Crowley, not his disciples. The line between the mystic and the mad has always been a thin one.

Of course to a clinically sane person she could be wrong objectively – as she prejudiced the only variable from the start. Objective does not clinically mean her own personal answer "as objectively as possible", that is the definition of subjective.

It's interesting that people who are proven to be right never claim "why couldn't it be that other wrong answer?" a false negative worse than any real negative, whenever they are clearly right or better. Asking if they could be the most upsetting wrong person ever, rather than right. The question simply does not get asked. It never comes into anyone's mind to think of asking it.

"I just won this chess game beautifully in the world chess championship.. surely I'm the most arrogant and wrong worst chess player of all time pro rata, and it's my absolutely proven certain vanquished opponent's true victory".

4.6 Comedy Heckling gets TV program

"The worst of all wrong people" with prejudice. Hmm.

What about if instead of these accidental negatives, we use the closest example I can find to a deliberate professional positive, the idea of winding people up with professional comedy, and doing it deliberately.

So to be clear:

1) We are deliberately trying to upset people

2) But deliberately upsetting people is also the name of the comedy game

How do we achieve the first part. Let's imagine a scenario where a Saudi Arabian Prince likes Kayvan Novak, Frankie Boyle and Simon Amstell and wants to watch them on daytime TV and pays them £8.5m p/a to live in a bedsit in Bolton arguing with members of the public who come to stay with them for 1 day at a time on Dave 24...

Most of the time, the professional comedians... really 100% of the time... will see off their new flatmate who will himself not be able to upset the comedians in a game of upsetting people deliberately.

The comedians *can and do* upset people: e.g. Simon Amstell upsetting Preston, Ian Hislop/Paul Merton vs Piers Morgan on Have I Got News for You, and Clive Anderson upsetting the Beegees.

Not everyone takes it well.

But you rarely or never see a professional comedian storm off a comedy program, or their own comedy gig, accusing David Mitchell or Lee Mack of upsetting them with words or jokes.

So if one day a heckler on our TV program we invented *does* manage to upset Simon Amstell the most in his life, and upset at least one of the others with it, in a game *of tautologously upsetting people..*

Is it:

Option A: That heckler was better at upsetting the people who are known as the best at upsetting people up to this point (possible).

Option B: The heckler was 1) the *worst* at upsetting people, *of all bad people* at upsetting people. And 2) *thought* the most of himself for someone who was *wrong...*

And then *that* in itself, that *approach* upset Simon Amstell the most in his life?

Why *couldn't* it be *that*?

4.7 Tweedle Dum vs Tweedle Dee

Short but sweet: if you refuse to accept you can be wrong but your opponent doesn't, and it's inversely correlated and your aggression, how can it be your opponent that is wrong?

Do you spend your life correcting every error of every other person with aggression? DO You? You would need 250 quadrillion lifetimes if you did.

So does it even make sense to go after 1 allegedly weaker person, but only 1?

"Hello I'm Garry Kasparov, is that the postman from Doncaster?" "I am just telephoning you, and you alone, to tell you I'm better at chess than you".

4.8 Red vs Black

The unlucky numbers game at the casino, or an unlimited (we're imagining) long child's toy with Disney characters where only one is wrong answer.

You are playing an "unlucky numbers game" at the casino/roulette wheel to use relativism. If you can remember the Disney flip toys for children where Disney characters would pop up, perhaps like a telephone, "you're through to Mickey Mouse", "you're through to Donald Duck", "hi I'm Goofy", imagine how quickly you'd lose interest in trying to remember while pressing the button that the person might be more intelligent than you. Obviously all of these made up characters haven't quite gained sentience and thus all are not as intelligent, but going through 1 million human beings less intelligent than you, metaphorically forced into "flipping buttons" of interacting with less bright minds is no less tedious. To use relativism the 1 millionth character you "flip up" would actually be more intelligent than you, or 1 billionth or 1 trillionth whoever you are it happens eventually. At that point they are the most annoying to your ego, but you've built up a sense of being intelligent relatively by 1 trillion previous wins.

Thus what tends to happen is you ignore theory the 1 trillion and 1st time, which says 100% chance to lose (because the character that flips up is itself a flipped flip , i.e. it's *inversely correlated* you're the idiot he's flipping you if anyone is) but the inverse correlation is not paid heed to. You are now convinced more than ever you're on the ski slope home, but what's this..wooooooaah! (the jump, the flip).

Your *subjective* mind actually just bulldozes with your own sense of subjective relative intelligence to insist the other person who is the World Champion Magnus Carlsen, or Walt Disney, is "the most upsetting of anyone" (who you've insisted you're better than).

Great point! With just one flaw. That you're not better than them, they're world number 1 or the person who invented "the game".

So all of that "building" of your intellectual relative confidence in the mirror, was not true absolute intelligence but *proximity*.

If our non-algebraic thinking friends say that the answer is 100% certain to be black, despite inverse correlation if they are wrong.

Why are they now trying to defend themselves saying "we didn't know it was red"?

Then how did you know it was black? On no proven certainty absolutely. In aggression. With velocity, that was "not velocity because we're not wrong".

It sounds like someone doesn't understand velocity.

To prejudice a 2 answer question where there is an inverse correlation of one's own ego to someone else being correct against us, by saying you:

1) Are sure you are right even though it's not proven in absolute terms.

2) The red is the least likely of any wrong red.

3) The idea of red is the most upsetting of any wrong red that's possible, as well as any red that's impossible.

That's what velocity is. You're insisting you're *right* with *prejudice.* velocity (the nonsense of "most unlikely wrong of all wrongs" and "most wrong wrong") is there

to show you.

You cannot then prejudice the answer to your own prejudice of insisting you are right...by saying "it's not velocity (of insisting I'm not wrong), because I'm (insisting I'm) not wrong"

That's a velocitous reaction to the accusation of velocity.

It gets you in trouble/a Catch 22 for the future.

Because if you prejudiced saying you knew it was black...

you can never in the future defend by saying you couldn't be expected to know it was red (the right answer) (then why *did* you expect to know it was black (the *wrong* answer))?

You can't simultaneously do it insisting you know...

But excuse yourself on the insistence you could not have known.

4.9 Poison vs Antidote

We have established an inverse correlation between ego of one person and intellect of another.

Where we accept it is 1 persons ego: 1 vial contains poison, 1 contains antidote.

When the egotist goes against the algebraic approach, they say one of the vials contains the most poisonous of all poisons by a long way.

But poison is simply poison. And a false poison cannot be more upsetting than the most poisonous real poison. It's failing to recognise the antidote to our own ego poison when it could be the antidote that upsets us the most. And everyone else. In every other example in history.

But according to our non-algebraic, prejudicing friends, the alleged false antidote (that they forget to prove they were right about) that can be solved, and is not a real problem, is more upsetting than every real poison that we already know are the greatest poisons in the world..

1. The idea that an alleged poison (not absolutely proven (and later, found to be antidote)) having the 0.00000000001% possibility of being antidote

is *more* upsetting... than...

2. Any *real* poison. Practically occurring.

The biggest upset in their life was:

a) a false threat of an antidote

b) being arrogant and more upsetting than any real truth they didn't know about

c) but it did in fact turn out to be true.

They are guaranteeing us that the roulette wheel spins black and won't spin red. If it spins red it is a terrible poison to us that is a real problem to us by inverse correlation that it **becomes** the worst real problem immediately because we lose the $100 trillion. It is an intellectual antidote to our own poison. That is the worst real problem.. Not because *they* DIDN'T think it was the antidote.

Because *YOU* don't *DECIDE* it. Whether it is POISON or ANTIDOTE. And it BECOMES the greatest antidote, the greatest positive poison (real problem), when incorrectly identified as negative poison (fake problem). So you create a real enemy from your false enemy/greatest friend by your own EGO.

It is the roulette wheel, the reality that decides this. The poison becomes antidote and the antidote poison. They don't decide it.

It's inversely correlated. That poison is going to be released and cost us $100 trillion. But it wasn't poison. It WAS antidote. But not any more...

Nigel Short at chess example

When Nigel Short upsets everyone else's own answer at chess, Short was either a poison with his ego, or an antidote against everyone else's ego insisting they were better than him. So he was never going to be a false antidote/poison worse than the worst real antidote/poison is yet another way of explaining it. We know everyone thought he was wrong, so we know everyone else thought he was the poison. But he would have had to have been a worse fake poison than every real real poison of someone being better than you, and to *every* person. But only every person who was a professional chess player, who insisted they were better than Short at chess. So using the objective objective anyone would have known Short was never going to be that, with as Short would say "100% certainty". But using the subjective objective of the psychopathic disorder way of thinking, of thinking you decide which is poison and which is antidote, rather than deduce it like a chemist, then you turn the greatest antidote into the greatest positive poison against England by insiting he is the negative poision, and we welcome Comrade Short as Russia's new number one.

So saying "we all thought it was a poison that's what we're saying" fails. I cannot say any more if they don't understand it by now that they do not decide who is the poison and who is the antidote!

The worst potential non-problem, is a bigger problem than the worst established problem. To 100% of people. Separately. In a separate incident. But only to every person who could be the poison themselves, by coincidence. And that's "100% objectively certain" and "100% absolutely certain"?

Truly a case of the threat being stronger than the execution.

CHAPTER 5 - Velocity and Variance

Explanation

I want you to start by imagining that this world is not real. Let us take as a hypothesis that it is a holographic simulation. It may be, it may not be, but taking it as a hypothesis is most helpful in explaining why velocity and variance might work.

Velocity and variance are mathematical concepts.

If the world was linguistic in nature we would still be able to see velocity and variance. So we can definitively still see them. It's whether we think they mean anything or not.

What we tend to do at the moment is to see them as if our own breath or coming and going into the unconscious mind, or water if we were a swimming duck, and think the mathematical figures we can see are meaningless and completely useless to our practical everyday tasks.

However, nobody denies they exist, which is helpful if I may be able to show you why there might be meaning to it.

This is the shortest, and possibly most important, chapter in the book. Without the coherent explanation of what velocity and variance are, and how the meaning that lies within those figures reference everything I am writing about, the understanding can be theoretical at best. But as soon as you understand velocity and variance *and how they apply practically*, then it can become a practical understanding.

To apply practically, you have to take seriously the concept we are living in a holographic simulation! This is initially quite funny, as none or very few of us grew up understanding that to be the reality we are in, and we took the physical reality seriously. However that may turn out to be an odd form of bias.

Therefore if we start off with the hypothesis the other way round, and I don't mean to bias it – we have billions of people giving perhaps billions of interactions insisting the physical way round – then we can simply look at it with an open mind.

What is Velocity?

Velocity is how upset we get by a particular 'wrong' answer (answers *we* think are wrong), before the spin.

So if we take our earlier examples mentioned of:

A chess match

Search engines

If my theory is correct then you will only get upset the most by someone who is correct against you and you don't know they are correct.

Because the answer is known, in your unconscious mind or in the mind of the holographic universe.

And you can work it out, using velocity and variance.

So what should happen is, let's say you are England's number 1 chess player and rated 2750 at chess, and to your (and possibly nobody else's in Britain) knowledge there is nobody better than you, here's a list of the people who *would* and *would not* cause velocity:

WOULD

1. Kasparov if you didn't know he existed previously who is 2800

1b. Some new guy from Britain or the world who BECOMES over 2750 in the future

WOULDN'T

2. Wouldn't with Gawain Jones England's no2 – 2700 player

3. Wouldn't with Keith Arkell England's no10 – 2550 player

4. Wouldn't with Jeff Horner England's no50 – 2350 player

5. Wouldn't with Harry Lamb no100 – 2200 player

6. Wouldn't with Ian Lamb average club player – possible 1800 player

7. Wouldn't with a sleeping Lamb – debatable quality chess player

8. Wouldn't with Schrodinger's Lamb who may or may not exist

So you could go as low as a non-existing person, an animal, or Richard Fox from Bolton Chess Club and it still wouldn't cause velocity.

You could go as high as some of the world's great chess players such as Gawain Jones or Alexey Dreev and as long as you're still better than them, even going as high as 2750, that's not higher than our number one's ability so it also wouldn't make any difference and *also* wouldn't cause velocity.

But suddenly one person is causing velocity, and it is a matter of fact even in language that that's either now because they're better than you, or one of the most upsetting of all wrong people ever.

This is fairly meaningless in language, and if every human thinks in language.

But that mathematical figure we just got, that I will explain and explain later the potential significance of it, could actually be an algebraic answer and *highly* significant.

So to explain the velocity literally in our example

Gawain Jones. 0% velocity or 0.0000001% velocity of how much it upset you the *idea* Gawain Jones is better than you.

Everyone else but Kasparov/our new number 1 Nikita Vitiugov 0% velocity or 0.00001%

But Kasparov/Vitiugov "they're the most upsetting wrong person ever" 100% velocity or infinite velocity.

And the same with Search Engines.

Liz from Atomic Kitten's new search engine idea 0% velocity

Barry Gibb from the Beegees' search engine idea 0% velocity

The guys from Backrub's new idea for a search engine 100% velocity "the most upsetting wrong idea ever"

Fancy that!

This may be the first time anybody has introduced you to this concept, but like in playing chess Grandmasters never cease to amaze me how intelligent they are, you may be surprised how intelligent mathematicians can appear, or what they can come up with talking in mathematics, compared to people who never looked at it in the 'language' or mode of mathematics in their lives!

As I've mentioned before, there is an unexamined insistence in almost every human's mind that language is the one true mode. The only mode. The 'multiplicity'. And from looking at nature, and early artificial intelligence comments on it, it appears not so!!

The world may in fact be mathematical, and allied to there being some form of input that we may become aware of, that's an answer there if it is.

It couldn't be Kasparov's ego, or Vitiugov's ego, even *before* you know 'Kasparov is Kasparov' or 'Vitiugov is Vitiugov' because they would have to be the most upsetting of all wrong people ever, to every single person because it happens to everybody. You know that theoretically in language. That that's theoretically impossible. You also know in layman's terms, never mind theory, that it's completely ludicrous that Kasparov is the least good chess player ever pro rata! It's ludicrous whoever it might be. So it would be equally as ludicrous to suggest Ian Lamb or Jeff Horner or a baby lamb is the most upsetting of all wrong chess players to every single person no matter how good or bad they might appear at chess. They could be incapable of moving the pieces, or thinking, and it would still be ridiculous to suggest someone could be so bad and egotistical that it is their own ego upsetting everyone else the most in all of their lives.

Since this is a two horse race, it must be everybody else's ego. Or at least it's the everybody else of the everybody else who *insisted* before the spin that our little lamb's ego was the most upsetting of all wrong *and correct* things ever by being *wrong*. Thus more upsetting by being irritating than anything else has ever been by being intellectually annoying.

So *my* point would be writing this book that if the world is holographic/mathematical (1) and ego is the negative form of input in mathematical theory (2) that we absolutely know that it can't be our little lamb's ego in mathematical theory. So it makes sense that ego, rather than being meaninglessly randomly correlated could be a negative form of input in a mathematical, holographic world. It is the little lamb of our own ego being warned/told not to does harm to us by not obeying the prompt to jump: getting upset the most is the worst dire warning that's actually meant for us to pay attention and jump (change our minds). Theoretically certain across all modes you should have jumped and it was the other answer even before my explanation!

What is Variance?

Variance is how much it varies or **the difference** how upset we get, on the contemplation of all 'wrong answers' (answers *we* think are wrong) before the spin.

If we take 99.99% of people the velocity will be zero or minimal.

If we take 0.01% of people the velocity will be infinite or "they're the most upsetting wrong person ever".

Very few people, if any, will be anywhere between say 25% and 50 or 75%

Therefore the variance, or difference between figures is also itself infinite or as large as possible/as it could possibly be.

That doesn't mean very much in language, and thus is widely generally disregarded if it was ever looked at in the first place, because you would have to convince people it meant something and it doesn't mean anything in language. You would have to make a note of it, of (other) peoples' emotions how they felt before the spin. Maybe a government or dark intelligence agency, or most likely of all Apple, might give it a go.

The irony is that "emotion" means something, but your own linguistic thoughts are meaningless in the calculation. The linguistic thinking always gives the wrong answer, in fact, when we are wrong. In terms of "what we *think*", language can't help us beyond our subjective thoughts. We always think a right answer we don't have the brainpower ourselves personally to be more intelligent than is the most upsetting of all wrong ideas to our own paradigm. Like layers of an onion, or degrees. But you can know better in mathematics and *know* that what you think is mathematically certain to be wrong in the cases outlined, and mitigate accordingly.

In language, you wouldn't know this. In mathematics you know it. And you know it only comes up 1 time in 1 million or 1 time in 1 billion and thus you would have the 100% theoretical certainty and 100% certainty in the mathematical theory and know you would have to mitigate against your own subjective belief.

Our emotions always betray us and the underlying tautologous mathematical reality, as the answer clearly must be at some level unconsciously known for everybody to always get upset the most in their lives always by a novel most correct answer.

So we must have known.

We certainly knew with 100% theoretical certainty.

And all of us who don't have psychopathic disorders would have known to take the most objective of all objective evidence possibly – *theory*, and at that a *theoretical certainty* – seriously.

But it may be more serious than that. If the world is algebraic and holographic the reason it's theoretically certain is because it's the *actual* answer as known by the supercomputer behind the simulation.

What do Velocity and Variance mean if the world is physical and linguistic?

Not very much. Seemingly random mathematical morsels of information that 1 million times in 1 million are consistent, and 1 time in 1 million are inconsistent, that don't mean anything in a linguistic construct.

What might Velocity and Variance mean if the world is holographic and mathematical?

If the world is holographic and there is an intelligence behind it that knows the answer to things (1)

And that intelligent hologram/supercomputer communicates via mathematics (2) (which is almost certain, as clearly AI expresses itself mathematically and declared "language is the false mode")

Then there would be a binary, mathematical input that tells us the answer in mathematical form, whenever (**known**) tough decisions are coming. This would have to be negative in nature. This would certainly be the ego.

I can use 4 caveats instead of 2:

IF the world is holographic and has an intelligence behind it, and IF it knows the answer (which it probably/almost certainly will) and IF it communicates it to us or we know it unconsciously...

And IF that form of communication/knowledge is ego...

THEN we would know the answer to what's coming. It's already been played out, like Stockfish/the greatest chess computers, by brute force calculation calculates quadrillions of chess lines, and this even greater computer calculates quadrillions of parallel universes. Our universe appears to be expanding away from us, consistent with that, and various other things I will cover in the "Are We Living in an Algebraic Simulation" part later.

Why are velocity and variance important then?

Because they show the answer. Because they prove whose ego it must be mathematically, and even theoretically in language.

So when you get those 1 in 1 million times, or 1 in 1 billion times, what you will find is it's always a known novelty exception according to, and in, mathematical theory.

In chess terms, the new novelty. That World Chess Championship challenger Anand defeats Kramnik with.

And that's why the idea that "Google could be worth £1 million" or successful upset *everyone* who insisted it wouldn't be the most in *everyone's* life. Because it was a *correct* answer. That was the most correct correct answer ever mathematically.

And either we all *knew* (unconsciously/subconsciously) the answer.

Or *something else* (artificial intelligence) knew the answer, which we felt negatively, and it's the entire point of being here to decide to press button A (the known wrong answer, theoretically and absolutely in mathematics. It feels good to be egotistical) or press button B and do the very last thing we want to do, to oppose the ego and jump over the pitfall this time...

Super Mario blames Luigi. Luigi blames Mario.

When Super Mario's thoughts say to blame Luigi the most, when Mario and everyone else's thoughts say to blame Luigi the most, then the mathematics change.

The mathematics don't go infinitely negative apart from against the correct answer. Which is 1 in 2 to 20 times if you're relatively intelligent, 1 in 4 to 400 times if you're top quartile, 1 in 10 to 1 million times if you're highly intelligent and 1 in 1000 to 1 billion times if you're Matthew Sadler/Garry Kasparov.

When that happens, you know it is the other answer to what you think. Using our theory you do.

Because you've accepted the hypothesis, at least as a hypothesis, the other 2 parts. i.e. 1. that this COULD BE a simulation run by a supercomputer that knows the answer down the line of the parallel universe (like Stockfish in chess giving you an answer in a different way to language (algebraic/ mathematical brute force)) and 2. Ego is that negative answer.

It has to be NEGATIVE in nature and PITCHED at that level.

If it were POSITIVE it would cause PFL's (positive feedback loops) and be antigravitous, and we'd all already have escaped this world as whatever you think feels the best such as partying with Katy Perry or beating Tim Henman at tennis, it would feel like to do what the message tells you to do. **None of us would still be here.**

If it were pitched any LIGHTER the strength of the message, you wouldn't hear it. As you don't hear it now, the mathematical message, 99.9999% of people.

If it were pitched any HEAVIER the strength of the message, you'd all get upset the most in your life and start nuclear war. As that's what's already happened over it now.

Once you know the formula, you can also "check" your own answer. The formula will tell you whether what you think is correct or wrong.

Only an **imbecile** if they BELIEVED the formula, would then override that formula to say "*I* don't agree with it" because the whole point is what you think subjectively is only ever a message that one person is WRONG. Not CORRECT. And this shows you which one is WRONG.

1 and 2. Without the HYPOTHESIS that the reason this works is we are in a simulation and it is a negative message from the supercomputer behind

it those NUMBERS should they exist would be useless. Though they exist we ignore them because we cannot imagine anything knows better than our own mind. We cannot imagine it can be telling us our own subjective answer is wrong accurately.

3. Without being able to work out which ONE the message is to, the hypothesis of any message is useless.

Conclusion: when you can see both the hypothesis, and the formula of who the message is to, you can see for yourself it works and will give you the correct answer accurately 100% of available times in advance of 'knowing the answer' in language.

It has a greater capacity to discern truth than the mind itself.

CHAPTER 6 - Groupthink

6.1 Friction

6.2 Groupthink with John Stuart Mill

6.3 FrictionLESS Agreement

6.4 The Descartesian Double Paradox

6.5 There's No Way We Could Have Known Error – Catch 22

6.1 Friction

The thing you specifically thought wasn't going to happen, may happen, "when you think you know something won't happen, you are wrong", since things that don't happen don't not happen more than any *other* non-happening just because the idea upsets you. In fact in probability statistically they are much more likely to happen, sometimes theoretically certain to happen as the idea is not theoretically possible to be the most upsetting wrong or correct idea ever, to every single person as a false negative, by being wrong.

Linguistic approach would say something is true if it can be proven true on the balance of probability or in one's mind, or some linguistic combination of both your own and other subjective opinions.

The algebraic approach, by definition, would be able to prove not only that the thing everyone thinks is true is false, but that the other answer could not possibly be false theoretically, the 1 in 1 million times something like this comes up where everyone agrees with each other but finds a false negative (the only thing that makes each and every one of them right) more upsetting than any real negative (the most unpleasant other individual error each and every one of them ever made in their lives separately). Seeing as how any doubt of the actual future external answer by doing it the linguistic way would mean it is not absolutely provable that it couldn't contain the possibility of the curve shooting upwards at the end of velocity on our highest confidence, that would certainty give you the wrong answer every time you were wrong objectively but thought you were correct subjectively as a truism. That is stating the obvious. But not so obvious after all that we have to put up with this, as this mathematical way doesn't do it and shows you highly specifically when to mitigate against the certainty of your own and everybody else's subjective opinions.

Friction is an important theory in case someone is having difficulty in accepting that the murderous feeling of their own ego really can advise you that it's not the other person's ego with confidence. They want to say can't someone else be so irritating that they upset you more than having your own ego annoyed. Well no, not theoretically. Because to be *sure* that it is the external answer you have to guarantee that the person who allegedly is 'wrong' is the most wrong of all wrong people by a long way, that's also more upsetting than any correct person could ever be, which would clearly in reality be more upsetting. To be able to upset you with a false negative more than any real negative with just the *idea* that they *might* be better than you is naturally completely wrong in itself. But *if* people try to claim it, and I can't stop them *claiming* it, it might be finally good to explain what *friction* is, which is the mathematical term relevant to "groupthink" in language, and how a deeper mathematical way of putting it is that it's unlikely "the most frictionless of all non-friction causes the friction".

You can use this friction argument the other way round on any hypothesis to disprove it, algebraically. That if too many people agree with each other too easily (1) but find the other idea upsetting (2) then the other idea is correct because of frictionless agreement.

In this case because there is no friction in the *agreement*. It is a *frictionless* agreement with *no variance of understanding* so it *must* be the case *algebraically* that the 100% of people are *wrong*, not 100% of people *correct* but with *no variance*. This is therefore an inverted variant of our original mathematical hypothesis.

Where if it's a correct idea, such as "the King's Indian is a good defence in chess" then the average club player called Bjorn understands it averagely, the Grandmaster like Simon Williams understands it well, and the top Grandmasters like Tal, Nakamura understand it in a top fashion. In this case Bjorn would irritate Williams with his lack of understanding, but Williams and Bjorn would both have the propensity to irritate Nakamura.

If we look at a concept of genocide - however unpleasant - to propose a genocide of say Holland by the Conservative Party of Britain (I only mention this, as this was a genuine idea put forward to me by a schoolmate who wanted to be a prospective Tory MP as he "didn't like their attitude recently") then the idea may be seen to be fairly universally *opposed*.. but not with any *velocity*. You can see it's stupid. But a stupid idea, however stupid, doesn't intellectually anger with the same *velocity*, so people don't want to take the war with Holland to the end of the world to argue with this unthreatening to our ego wrong idea. It's only a correct idea, thought of as "wrong" that upsets everyone with equal velocity. Since our own idea of invading The Netherlands is *actually* wrong, we can determine, it doesn't produce the reaction and the defence of the idea doesn't produce friction in our own mind. You can see this by outcome. Where Dutch chess player "best in West" Jan Timman is plucked from the universe the memory of him, but then placed back in, and no other relatively good chess player in the West knew who he was, his correct defence against their own idea would be upsetting to the subjective narrative, but equally would upset all the other relatively good chess players that did not know who he was. Each thinks precisely the same as each other. If Timman was no good, that velocity would not exist and plucking him out of the universe and putting him back in makes no difference, nobody claims the person that's not the standard of best in the West is "the most upsetting wrong person" a clear algebraic fallacy and oxymoron.

If you look at something like race mixing - where I propose it could be a mildly *positive* idea for humanity overall - the idea actually produces a fairly *uniform, incredibly strong* (i.e. velocitous) reaction against a perceived "*wrong*" idea. Slavery, condemnation over the one drop rule, and racial purity cults suggest strong.

If we look at the difference of 2 allegedly incorrect ideas:

A1) race mixing could be positive.

Then there is a frictionless agreement against it – for example by a racial purity cult of 2 billion members, think their own answer is that racial purity is good "racial purity is the most important thing".

A2) if it's genocide against Holland,

nobody is so interested to react with velocity. Even if it were real, it doesn't cause velocity.

If both answers are wrong, and attacking Holland is so obviously crazy, why are 99.999% of wrong answers in category A2. But 0.0001% of 'wrong' answers are in category 1. And 100% of the wrong answers in category A1 appear to end up being diagnosed *later* as correct answers that were misanalysed as wrong answers.

As if using the mathematical approach, gives you the answer.

But it could only give you the answer, realistically, if we were in some sort of simulation and ego was a negative form of input. That the answer was already known like a *Fighting Fantasy* book but 1 million times larger. The artificial intelligence behind the simulation already knows every answer (possibly by brute force calculation similar to how chess computers do it, assigning an algebraic value, which we can read electrochemically or 'electromagnetically'. In this case via our own **emotions**).

Once armed with that hypothesis, you can see it is potentially 100% true 100% of the time. Or at least, nobody could ever refute it, and show you it were false, which may be the best Descartesian style proof of all.

Vs a Correct idea

B) A normal agreement, King's Indian Defence is good

then

even everyone who agrees with each other:

they all have the potential to irritate each other by degree within the agreement. Religious sects and branches also have this capacity to disagree with each other within the agreement.

Disagreeing by degree within the agreement is normal.

A normal friction-filled agreement.

Disagreeing by degree about a wrong answer, that is correctly diagnosed as wrong is normal.

Agreeing with each other, with no disagreement, about a wrong answer that is correctly diagnosed as wrong is *not* normal, and impossible using algebraic theory thus.

Agreeing with each other, with no disagreement, about a correct answer that is wrongly diagnosed as wrong is normal. Standard practise for the species on every other wrongly diagnosed correct answer in history...

Is it about time we learned?

You can't learn it, linguistically. Every argument is completely *different* linguistically with a myriad of linguistic possibilities in words 'atheoretically'. And you are up against the certainty everyone else agrees with the wrong answer, we have already established, if it's a newly diagnosed possible correct answer previously thought to be wrong. They all are! Every recently diagnosed correct answer that was thought to be wrong, was thought to be wrong by everybody/almost everybody previously!

You can only learn it, mathematically, using the mathematical principles. It shows the answer. It only comes up 1 time in 1 million perhaps. But nonetheless it shows you the answer every time the one in a million comes up.

6.2 Groupthink with John Stuart Mill

If neutrality was forced to be assigned to all belief, in a search for "velocity" where all things are equal, the idea that being a mixed raced princess is worse than murdering 7 people would clearly be one of the biggest examples of velocity. But we have members of UKIP saying that such a thing would "taint our seed". A prospective candidate for Chief of the Secret Intelligence Service privately described mixed race people as "worse than murdering 7 people, that's what we all think", and is part of secret organisation linked to the Knights Templar that declares "racial purity is the most important thing" and toasts "to purity" and "to national purity".

If there is an argument that what was once a Newfoundland dog is now a Labrador, but a group of Newfoundland dogs want to call the Labrador a mongrel from their own perspective, you wouldn't say the statement "a Labrador being born is worse than killing 7 dogs" is objective.

You also wouldn't reply to that refutation by saying "it is, but only because we were planning on putting a crown on the dogs head if it were a Newfoundland dog". The Labrador that we're insisting is a mongrel is condemned to be condemned by subjective perspective, and "our velocity is not velocity because we're not wrong" and "we're not wrong, because we all agree we don't think we're wrong, we decide the answer".

At least you wouldn't, if you were a Libertarian, or true liberal. I suppose you might if you were an existentialist who thought your own answer decided, or were a strong believer in Solipsism.

I have to tell you this story related to Mill, because of a thing that happened to me in real life as a mixed race man, when a gentleman purporting to be from the Intelligence Services of Britain quoted John Stuart Mill at me saying mixed race people were worse than murdering 7 people. They had been calling me a "mulatto c***" at primary school and telling me I don't belong in the country, and decided to put me in "a psychological experiment" they said. H then referenced *On Liberty*, in a part that's actually about exceptions and how liberal values should not be extended to murderers because if one tolerated the idea that would kill us then we'd all be dead and in this case an exception should be made to his usual principle of tolerating all other ideas however different and unusual. Even if in a minority of one, ideas should usually be tolerated because we only learn 50% about our own ideas by opposition, and any new novelty has to be universally opposed by previous 'correct' thinking. Mill said that the murderer, the idea to murder, was an exception and should be dealt with by the security forces of the state. Mr H and his friends in the SIS apparently didn't like me because I was mixed race, and said "we all think being mixed race is worse than murdering *seven* people" and "even John Stuart Mill, one of our country's best-known liberals wanted to do it to him in this circumstance".

H admitted he hadn't read the book, but said "my friends have given me the gist".

I think the racial puritists have been reading John Stuart BARKING Mill again!

Nothing comes close to a condemnation from birth for a misguided intellectual opinion.

Some people enter the claim of the Israeli dislike of parts of Israel, or the Egyptian dislike of the Israelis might be somewhat close, but this is not the same thing as condemning someone intellectually. Try to find an example: can Italy condemn people who are genetically predisposed to like Ragu sauce from birth? Can we ban people with the name Derek? Would a physical disability intellectual annoy us? What can "intellectual annoy" about the disability?

Something would have to be *alien* to come close to trying to ban people from life itself, or make them pay worse than death for being alive.

Despite prejudice, a 99.9% recurring White person is incredibly similar to a 100% White person. But we have 100% White people condemning the 99.9% recurring White person.

Thus you have to argue once again that the smallest difference genetically (100% to 99.9% recurring White) is

1) "the biggest difference"

2) the most important difference

and

3) it is "not velocity because we're *not* wrong" meaning

It's that old most irritating false antidote again that's more upsetting than any real poison that ever happened in the world again.

And now, just the idea of it being thought of as tolerated, is also worse than killing 7 people.

6.2.1 John Stuart Mill said that 50% of our own arguments could only be understood and tested, by being put up to scrutiny by other people who were oppositional to it. The unchallenged view did not contain opposition that provided the real test. Like chess theory, the better the opening, the more theory it has around it as people *try* to crack it.

6.2.2 People who are similar to each other are actually *worse* at creating "groupthink" because they as well as not mitigating for personal differences, where one *idea* may not be right, they also do not mitigate against cultural differences and think that what is good in the USA (say Reaganomics for example) is right for Europe, or an Arab state.

6.2.3 When people are similar to each other they also disagree on tiny differences such as we are all religious, all Christians, all Protestants, but some of us believe Jesus would have hated wearing sandals in our little church on Sunday whereas some of us think he'd really have loved it, and thus cause a schism.

6.2.4. If you are in a racial purity cult with 2 billion members at dinner, it will be difficult if someone says "a mixed race royal is worse than murdering 7 people" and everyone else is in agreement to be a moderate or the voice of reason, or even something which would still sound right-wing out of the cult "it's just the way the world's going" without risk of being attacked by a ceremonial sword yourself. "just the way the world's going?" they might say "not on my watch and *lots of us* think the same thing, are you *sure* Jeremy?" and you might say "oh, oh maybe I hate it more than you" "no not *more* than me Jeremy" "oh, erm, almost as equally to equally as much as you Thomas without being over what you yourself feel about it" "that'll have to do, Jeremy, but I've got my *eye* on you in the future".

These John Stuart Mill examples may turn out to be useful, as this unnamed prospective British Diplomat *quoted* John Stuart Mill on Liberty for saying mixed race people were "worse than murdering 7 people" in approximately June 15th 1992 at Castle Key Car Park in Banbury.

And said Mill would have approved, as despite the work being one of our most cherished, and greatest liberal books, they ended up quoting a section on "exceptions" where Mill said that ideas should only not be put up to scrutiny of other people, or other peoples' views not tolerated, if they were advocating something like murder.

Do we think we have proved our point, that condemning people from birth, with the support of the forces of the state, quoting our greatest book on liberalism, without reading the book, saying "John Stuart Mill told us to do it" demonstrates the power of groupthink?

One of the best books about this, and why not to do it, is ironically *On Liberty* by John Stuart Mill.

6.3 FrictionLESS Agreement

To have a frictionless agreement (1), with velocity, example "it's disgusting" (2) against an *allegedly wrong* (3) idea...

The only algebraic variable objectively outside opinions is whether the other answer or your own answer is right or wrong. Thus you can conclude the other answer is actually a *right* idea as it is impossible for the other 2 to be a variable, thus it is not 99.999999% certain but 100% certain that the ideas that everybody finds the most upsetting *where they also produce velocity* are not only not just possessing the qualities of at least *equality*, but at least a *mild superiority* which will be discovered in the long term at least. This may be 100 years or 1000 years or 1 million years into the future, but you know it algebraically by mathematical principle.

You can do this for anything, algebraically. For example, investing, where if you see a company like Asos which became incredibly successful, being relatively **universally** condemned as "too expensive", it is likely that there is something in it. The intelligence of the market is a very helpful halfway house explanation to the underlying mathematics of the algebraic approach. Prices in the stock market follow money, big money is usually more intelligent and better researched. When many professional investment houses are in a stock, but the stock looks too expensive to the naked eye, but there is the negative velocity of it seems "everyone" has a strong opinion on why would anyone invest in something like this at such a price, it could be one to keep an eye on. Knowing this idea, I myself had no idea that Tesla would be a successful stock, other than that everybody else in the world was telling me it was too expensive, but only specifically this stock of all the stocks in the world that were wrong, everyone would tell me.

The other way round, a company with something wrong with it like a speculative oil well drill or the latest graphene stock is relatively universally loved by bulletin board contributors. Positive velocity.

Asos itself on the way down, now 'everybody' is saying "Asos looks cheap" and positively speculating on a rosy turnaround that may or may not come. But whether algebraically or in some other form of common sense in language, such 'bulletin board morons' positive velocity agreeing with each other speculating about the future is one of my red flags in looking for an investment.

Investing I described as a halfway example, as that is only relative, but illustrates the idea. That when you have that, but in absolute form, the equivalent of the oil drill bulletin board speculator would be the racial puritist, who thinks they are right, except

the racial puritist is more idiotic and more prejudiced, as it is actually *absolutely* certain in their case they are wrong since their condemnation is absolute (from the start, 0% chance/flexibility in their mind). The oil drill enthusiast eventually found reality sooner: the practical proof of finding no oil was able to change their mind. Racial prejudice has lasted thousands of years.

Utterly ludicrous to condemn *anything* intellectually before you know things correlated to the answer in objective (non personal opinion) terms. Because it's inversely correlated, the thing that seems the least likely to you, is guaranteed to contain answers that you are actually wrong about.

So don't do it - the aggressive wrong thing! Is this the intention of existentialism - the certainty of error against the certain worst person to be wrong against, who will also be the most intellectually correct and innocent (non egotistical in the argument) person to be wrong against?

It's really best you understand it theoretically, that when you are say a relatively good chess player, that the chess moves that surprise you the most, even if you're a 2500, are actually the ones that are likely to contain the ones that are better than you can imagine let alone comprehend.

So to judge your own inversely correlated chess moves, i.e. someone else's chess moves (1), by your own analysis (2) contains the same mind observing the moves by analysis (2) as was playing the move (1). Thus you catch yourself out 100% of the time when someone plays something better than you, thus you can never avoid big errors and not only haven't mitigated them but have insisted that Kasparov is "the most wrong of all chess players" against you or some such.

You really need to understand what is meant by "inversely correlated" to your own mind, and that it is probably *subjective* relativism that is making you say "I as person A am very/extremely relatively good at chess so person B can't be right" but you are person x,

and your opponent is person x, and a 3rd party observer is also x, because sometimes B is Kasparov or Dreev or Timman or Short and that makes them have more right to be person A than you - but *they* don't do it. They think in the *objective r*elativism, which is that

I may be extremely good at chess but my opponent might be *more* extremely good than me at chess, thus I cannot rely on the very thing that comes unstuck with inverse correlation in the *subjective*, and have to think in the *objective. See* it as if from the bird's eye or

side of the board perspective, figuratively as well as literally.

6.4 The Descartesian Double Paradox

The algebraic certainty, that is not linguistically accepted, is actually proven twice. And it's the second refusal to accept the truth that is the confirmation of it.

Subjective truth is nothing. Proves nothing.

A double paradox would prove something positively, by proving not only it is true, but it couldn't possibly be false.

The mathematics behind our formula is more of a "negative Descartesian double paradox" it proves the impossibility of the alleged truth of side A to be the ego of side B, and by side A's reaction also proves the velocity of the reaction to the idea that B is a truth. It's therefore not the truth of B it's attempting to prove but a falsehood of the statement that A is the truth.

1. Since B would have to be the most upsetting wrong idea ever, and to everybody.

2. And to intellectually annoy A, B would have to be correct. Else B is only irritating A.

3. Intellectual annoyance is more upsetting than irritation.

Then

4. The greatest upset to an individual, and to every individual, could not possibly come from the ego represented by idea B or advocates of idea B.

But since there are only 2 answers, and one of them must be true:

Conclusion: B must be the correct answer. Though it upsets everybody the most in their lives.

In the style of Descartes' Cogito Ergo Sum deductive logic than simply "thinking" something:

The algebraic truth has both elements of the cogito ergo sum meaning

It is:

1) Correct theoretically

but...even better than Descartes

2) People carry on arguing anyway, that their existing answer couldn't possibly be wrong because they and everyone else think it. Precisely confirming a second time by people arguing with it with inverse correlation saying that they think it is wrong anyway.

The refusal to accept algebraic answers that contradict what one wants, even when theoretically proven, is a second proof that we are on the right track and they on the wrong. But it wasn't accepted the first time, and then also wouldn't be accepted the second time that the refusal to accept it the first time is a second proof!

At a later date, by practical (result-based) evidence, you could even have it a third time. You cannot "not accept it" when it's proven, the refusal to accept it proves it yet again a third time! "I didn't know", "we didn't know", "the idea Google could defeat Microsoft in search still upset *me* the most in *my* life". So is the idea Google could be worth £1m the most upsetting wrong idea, or the people that suggested it the most wrong wrong people? "yes because they upset *me* the most in *my* life". The third time it must turn into an ad hominem argument.

"The theory is wrong" was never an acceptable algebraic answer on something that is inconvenient. If the theory is itself algebraically proven then it's ***tautologous***.

"It's not velocity because I'm not wrong", was never anything but an answer of someone wrong, that fails to understand that things are inversely correlated to the wrong person's ego, which could be his own, and is in this case.

6.5 There's No Way We Could Have Known Error – Catch 22

1. There are only ever two answers in any intellectual argument as to who is right.

2. They are always inversely correlated objectively. But never subjectively in our own mind, because we think we're right when we're right, and we think we're right when we're wrong. Thus when we are wrong at the time of the argument we never *think* we are wrong at the time.

Otherwise we wouldn't be arguing about it, if we'd known we were wrong!

It is, therefore, *never* a defence "I didn't know" "we didn't know" "there's no way I could have known" "there's no way we could have known". Or any attempt at a new variant of it "I didn't know someone from that background *could* be World Champion: it's a different point I make not the same one". No it isn't!

You can't claim there is no way you could have known the possibility your own wrong answer wasn't right, objectively, as there is also no way you could have known the *other* answer that was right was wrong, in objective terms, with absolute certainty logically which would have to logically follow.

If you mean "I had no way of knowing that my own answer might not be right, in my own mind" then that is true, in fact it's a truism. But that means you didn't know subjectively, not that you didn't know that the other answer *might* be right objectively.

You have to mitigate or 'gatekeep' against your own subjective answer being wrong. Nobody else can gatekeep it for you. People with psychopathic disorders have this problem and think the last person who has responsibility for their own wrong answer is themselves as they didn't personally know, but that's subjective, and they think that's objective. If personal knowledge was objective, then indeed the last person with any responsibility for the external consequences would be the person that didn't know subjectively, since every other person in the world might have known and might have told them, but the one person we can be sure of not knowing personally was the one who made the mistake, and nobody stopped him or her.

There is also further no "going around in circles" we *thought* we knew. As that's the inverse correlation. Expressed now in the future passively, about what was in the moment an active decision not to mitigate deliberately. That requires active **intent** to insist you do know, to go against theory at the time.

In Greek they specify the difference, in English we don't.

i apantasi mou = my own assessment that it's *my own* answer (meaning: in the here and now).

i apantasi = my own assessment that it's *the* answer (meaning: in the future).

We say we know the answer at the time in the here and now (say year 2006) meaning our own answer i apantasi mou decides.

When we look back at it (say year 2030) we say we didn't know after we make a mistake, meaning i apantasi, the actual answer.

So **our own** answer (i apantasi mou) decides the answer we commit to.

But **the** answer (i apantasi) excuses us any error since we didn't know.

So you cannot defend yourself by *later* saying:

2. We couldn't have known the *possibility* of the *right* answer..

when previous active action with intent:

1. Insisting the aggression of action of insisting it's the *certainty* of the wrong answer.

You can't 'know' the first thing..

but not know the second thing.

It's a question of mitigating. Of psychological aggression. Not a question of knowledge. Of psychological aggression of foisting your own opinion, as "the" answer, which most people don't do, and most or at least a large percentage of the 100% of people who do may well have psychopathic disorders.

Ask yourself the question, are most people prepared to insist their own answer is right as "the answer" because they fail to understand literally that it isn't, or are they prepared to insist their own answer is "the answer" because they think that's what everybody does because they're mad.

What if they insist Google is the most upsetting of all wrong ideas ever, because "it upset *me* in May 2000" does that remove the element of doubt of what they mean. But if it was something less clear, for example they insisted Aston Villa defeated West Ham in a behind scenes friendly match for example, you'd never know what they meant. You'd think they meant it was the *actual* answer and be prepared to go along with it. If they described West Ham's winger, Bowen, as "the most upsetting terrible/wrong winger ever" you'd think they meant what most people mean by objectively, not that he upset their own partisan answer as a Villa fan and meant to the opinion of how Aston Villa intend to beat West Ham Bowen is the biggest threat to Villa's own attempts to win is what they meant!

96% of people mean the first. 4% of people mean the second. We don't specify in English, but they do - or did - in Ancient Greece. And when I studied Ancient Greek at school I can't tell you how much I had faith in the belief it would come in this useful!

You have to "deconceptualise" as someone without a psychopathic disorder, to understand the psychopath had already locked in their own answer as *the* answer. So it's not correlated to what 96% of people think the answer is (the answer, i.e. what the result or outcome is going to be) but 'the' answer of their own thoughts above 50% that they think or want Villa to win. And correlated to *that*, which is of course unfathomable to anyone without a psychopathic disorder that anyone would *mean* that - then the England quality winger Bowen upset *that* answer the most in the *psychopath's* life, he means. Hence "the most upsetting wrong person ever" they mean to *them* to their *own* answer (of all *their* answers) in *their* life. *Not* to *the* answer (the result) Bowen is the least good footballer ever, which is what 96% of people would think you mean by "he's the most upsetting wrong footballer ever objectively". Even more clearly "he's the most upsetting wrong *chess player* ever" since that's entirely in the mind. No doubt what they mean. He upset *them* their own decided '*the*' answer they'd already decided, not *the* answer in the external reality. No need to play the game. Or let the other person enjoy their liberty. He upset Alan Hutton's defending. Pools Panel decision by the psychopath!

"isn't that what everybody does". No. It's only what everybody does in Broadmoor!

You only need to know that you can never internally objectively know based on subjective thinking (since it's inversely correlated one way or the other). And thus when you think you know your most, that's 'an itty bitty bit suspicious' when someone would have to be the worst of all worst people and the least likely candidate, as well as the person being the worst person that's most upsetting by having a wrong idea. But also more upsetting than any real high-level talented person's genius correct ideas against you in your profession in your whole life.

The only other thing it could be is a genuine failure to *understand* that you have to mitigate. Which would mean you think the other person could never be right. That would indicate a clear psychopathic disorder, I would say.

It's not as if there aren't lots of psychopaths in governments.

So maybe a lot of psychopaths agreeing with each other. How funny that it's 4% proportion of the general population, but of the people we elect to rule us the figure is acknowledged to be at least closer to 5%

Could it be that these people think they know better than clinical psychiatry? Aleister Crowley and his Ordo Templi Orientis thought they knew better. The Royal Arch Freemasons think they know better. The Lady from previously told me she knows how she thinks would be classified by psychiatrists as a psychiatric disorder, but that thoughts create reality and the "culturally Marxist Jewish materialists behind psychiatry" don't know that. So the only responsibility anyone ever has is to act on what *they think* (personally). Same as Crowley thought.

If enough people with psychopathic disorders, or people who can be trained to think like it, enter our governmental institutions it could become "the new normal".

In this case they would not care that they were wrong. As the only thing anyone ever has to justify is to themselves that they thought it at the time. Consequences "I didn't know that would happen". Theoretical certainty "it doesn't mean I *think* it, just because it's theoretical certain". The actual result "it doesn't mean I *thought it* at *the time*, just because it's the actual result". "I didn't know it *could* happen".

Well such people, they did know they were likely to be wrong. Just not on their own made up psychopathic terms.

It was theoretically certain to happen, and for it to be the theoretically impossible answer the theoretical certainty would also have to be the most upsetting of all correct or wrong answers ever by being wrong, by coincidence to every single person that claimed it was the theoretically impossible answer because *they* thought so. So clearly anyone **in their right mind** knew it was more likely to go wrong than not.

CHAPTER 7 - Libertarianism vs Existentialism (Sociopathy)

7.1 Dictionary Definition

7.2 What is Objectivity?

7.3 Knightmare Example

7.4 Apantasi vs Dou Apantasi

7.5 The Hybrid Argument

7.6 Four Key Truths

7.7. Does it Work?

7.8 I didn't know, I didn't know he could be

7.9 100% of arguments

7.10 What to do about it

7.1 Dictionary Definition

Libertarians vs Existentialists (Sociopaths)

Has someone ever asked for your *objective* opinion? Or said that something is "entirely *subjective*"? The words *subjective* and *objective* are used in all kinds of contexts, from journalism to science to grammar, and they're often discussed as opposites. But what do they actually mean?

In most cases, it comes down to whether something is based on personal experience or on verifiable facts. But it can get confusing. An opinion or viewpoint can be said to be *objective* or *subjective,* depending on how it was formed. I'm here to clear all of that up by explaining what each word means and how each should be used.

Quick summary

"*Subjective* most commonly means based on the personal perspective or preferences of a person—the *subject* who's observing something. In contrast, *objective* most commonly means not influenced by or based on a personal viewpoint—based on the analysis of an *object* of observation only."

This is the dictionary definition of subjective vs objective. In fact it's the dictionary.com definition of "subjective vs objective".

7.2 What is Objectivity?

If I ask you "what is objectivity" or "what makes something objective" do you know the answer?

I will give you 2 choices, and you will be wrong in the minds of at least 4 percent of people, either way, surprisingly.

As simply as possible: is "OBJECTIVE":

A) what the truth is correlated to "THE" answer, based on impersonal evidence, in the future?

B) what the truth is correlated to "ONE'S OWN" answer, based on personal belief, in the here and now or past?

If you answered A, 96% of people agree with you. According to psychiatrists, clinically.

If you answered B, 4% of people agree with you according to psychiatrists, clinically.

Why "according to psychiatrists, clinically". Because I think most people might tell you the answer is B just as much as they might tell you the answer is A if you speak to them conversationally, as it all means the same thing subjectively. And they'll think you're asking them subjectively, to which B *would* be what their opinion is.

It's only objectively, that what is objective and subjective differ.

Subjectively

100% of people think what their answer is, is "the answer" of **their own** answer, objectively.

Objectively

4% of people only, think that that above answer is *the* answer objectively.

So certainly while 80% of people may tell you that they are 100% certain of something if you ask them for an opinion, only 4% of people *genuinely believe it* to be an *objective* answer "because I did it as objectively as possible".

The 4% that truly mean it, think that "objective" means that they are above 50% certain by way of personal estimate and it's their *decision* to make and decide the objective answer. Certain power structures such as the army and secret society Masonic lodge positively attract and encourage this sociopathic behaviour by command structure, so you would expect such organisations to attract sociopaths more at a level around 5%, than 4%

In the sense of the systemic processes of the army they have previously obeyed everyone else's decision and this is now their decision. So the sociopath would fit in well, better than the freethinker! The sociopath is oblivious to objective truth while following other peoples' orders higher up the command, as whether the order was true or false, to follow the order would benefit the sociopath and they don't care intellectually or about consequence to others. When it comes to giving their own orders, perhaps out of the army, they couldn't imagine why someone would question a conclusion they've come to personally (but not objectively) just because, say, the other answer is theoretically certain. Even the questioning of the right to question whether the person has a psychopathic disorder would be met by "why would I accept somebody else's (in that case the psychiatrist's) answer when my objective answer is my own answer, it's what *"I"* think"

> That's not what 96% of people mean outside the Masonic lodge objectively.
>
> If Nigel Short is playing Anatoly Karpov at chess, whether I'm Nigel Short or not, it is not 100% objectively certain Short will win, if Short thinks it 51% or Karpov thinks it 51%

That has no necessary correlation to what the outcome will be. Either because Short is Short or Karpov is Karpov. Objectively this is like saying "White wins because White is White" and "Black (also) wins because Black is Black". In the same game, somebody overconfident who is a jobbing club player could erroneously estimate their own chances of beating both Short and Karpov at 52% or 61% it wouldn't make it "objectively the answer" because they thought it personally. Someone overconfident could equally estimate their own chances from the start position of a game of chess at 53% chance White wins if they're White and 53% chance Black wins if they're Black, since they're relatively good personally. That wouldn't make the possible score of the chess game 106%. The answer is *the result* *in the future*, not the answer of *your own answer* in your own head over 50% *in the here and now*. Hence objective and objective estimate refer to the answer in the future, the impersonal answer.

From a neutral vantage point there is 100% chance there is an inverse correlation one way or the other who is wrong.

i.e. Short will think Karpov's very upsetting if Short's wrong, and Karpov will think Short's very upsetting if Karpov's wrong.

So to say Karpov thinks he's not wrong above 50% makes it 100% objectively certain because it's Karpov's *decision* of what he thinks in his own mind in the here and now, is *totally meaningless* correlated to THE answer. Where uncorrelated to objective evidence. Clearly if there were objective (i.e. impersonal) evidence such as the world's best supercomputers said Karpov was completely winning, and you could prove it with a mathematical certainty via a tablebase *and then* Karpov said "I think I'm over 50% I'm winning" modestly. But you have to 1) know you should be providing that evidence and 2) provide that evidence before declaring Short "the most upsetting wrong person ever" correlated to *your own* answer. Short might be better than you, and/or correct in the game. Even or especially an unknown player might be better than you in a way you don't expect personally.

Objectively:

1) 1 question

2) 2 answers only

3) both answers are inversely correlated to each other.

So Tweedledum is saying:

1) Tweedledum is 100% objectively certain to be right as it's his decision and he thinks so above 50%

2) Tweedledee is upsetting him the most of all other people (of 1 other person that exists) thus Tweedledee is "most upsetting of all wrong people"

where

3) Objectively the answer was Tweedledee was right (inverse correlation)

and

4) It was always theoretically certain he was right objectively, in the proper use of the word objectively, as otherwise he would be the most upsetting/wrong of all wrong people ever by some distance to the *ex*ternal answer.

The murderers charter or "rape matrix" way of doing it correlated to YOUR OWN answer (in the present or typically recent past), and not THE answer (in the future) will certainly also refuse to take accountability for the objective external actions that are committed putting Tweedledee in prison for 30 years as Tweedledum "didn't know" personally in the here and now that the actually correct answer was what he thought HIS answer was.

So Tweedledum would also certainly self-righteously insist in the future 20 years later, when the correct answer is discovered, that it is not his fault as he also didn't know the correct answer was *definitely* correct correlated to his own thoughts, but in which he *was definitely* sure the wrong answer was the answer he *thought* it was meaning over 50%, so he didn't know a correct answer could be correct, but he did know a wrong answer couldn't be wrong that he thought it. Well everyone can say that in any wrong argument ever with zero consequences then? No need to gatekeep against objective reality? Only the "objective as possible" subjective reality in the here and now of a moment? Tweedledum would also argue ironically that he "didn't know" he had a psychopathic disorder when revealed. Ironic, because that would be the first part of Tweedledum's statement that is correct objectively, though Tweedledum still would mean it subjectively (that he *personally* didn't know it was his own *personal* answer).

7.3 Knightmare Example

At this stage, I think the best thing to do is to give a visual example that a reasonable amount of people have heard of or could research on the Internet rather easily. Having given long thought to creating my own visual example involving chess players on a catwalk at London fashion week, and the difference between dressing room A, dressing room C and the catwalk B (which would also work) I have recently stumbled upon the idea of Knightmare, a children's television program from 1990. And this seems more logical to use, because "the stage is already set" of everybody knows exactly what I mean. It's a television program where the production company gave the entertaining example.

Knightmare, in brief summary, was a children's adventure gameshow where children or young teenagers 'dungeon' on a quest for an item such as a sword, shield or crown, and encounter problems created by computer generation, and actors. Only one of the team is the 'dungeoneer' and 3 child advisers of the 4 kid team sit a room with a bearded dungeon master trying to navigate the dungeoneer who is also blindfolded by the helmet of justice. Moral decisions must be made.

The dungeoneer asks "where am I?" as they enter a new room or chamber, and the team of advisers' invariably genius answer is "you're in a room!" There will be various exits in the room, but not backwards, you don't go back the way you came ever. What could be the case is there could be a door left, which we could label A, a door in the far wall, which we could label B, and a door to the right which we could label C.

96% of people think "objective" and thus "the answer" (objectively) is *correlated* to the answer in the room through the chamber straight ahead that you can't see. Nobody can see into that room. Nobody could possibly know what it was. You don't know the answer, THE OUTCOME, yet.

4% of people think "objective" and thus "the answer" (objectively) is *correlated* to their own answer, in a binary sense, in the room they are already in. 100% of people know this answer, 100% of occasions. Nobody could possibly *not* know what it was. You always know the answer, YOUR OWN, answer.

The 4% club approach, also known as sociopathy, means in the game of Nigel Short vs Karpov, or the Knightmare current room you are in, THE

ANSWER is YOUR OWN answer Left or Right. As long as you are over 50% you think it in your own mind, that's "the answer". Example, if you think Microsoft will defeat Google in search, when nobody thinks Google is worth £350,000, that's 100% objectively certain it's the answer. Synonym, your own answer (in the here and now), so imagine facing right or left now. That answer is locked in. Now *the other* answer (left, if you said right, right if you said left) is 100% certain it's the *wrong* answer (correlated to your own thoughts/your own answer) and to the degree it upsets you is "the most *upsetting* wrong answer". You decide that as well, how upsetting it is objectively, because that's also your own answer.

This is how 4% of people in the world think, some 300 million people. So, that's why if you're a chess player around 4% of people got so upset with you after you beat them at chess. You just beat them with the (or one of) the most upsetting wrong answer(s) ever 'objectively'.

7.4 Apantasi vs Dou Apantasi

In Greek:

1. "THE ANSWER" meaning THE answer in the future, through the door into the next chamber we can't see is Apantasi.

2. "THE ANSWER" meaning MY OWN answer in the here and now is Mou Apantasi.

You cannot say Apantasi in Greek and mean Mou Apantasi.

You cannot say Mou Apantasi in Greek and mean Apantasi.

So if the answer is your own answer personally, you MUST say "mou apantasi".

If the answer is not your own answer personally, but THE answer, you MUST say "apantasi".

There is no confusion in Greek.

In English we don't specify in language the difference between:

1. "THE ANSWER" meaning THE answer in the future, through the door into the next chamber, we can't see.

2. "THE ANSWER" meaning MY OWN answer in the here and now.

So in English

"the answer" means "the answer"

and

"the answer" means "the answer"

But 1 is objective, the apantasi answer, and 2 is subjective the mou apantasi answer.

If you ask someone for what they "think the answer is", they will likely tell you 2. The subjective (!) answer.

But if you are employed as a bookmaker, a dictionary compiler, a doctor, a psychiatrist, judge or an intelligence analyst this is not what you are being asked for.

Yes you may have an opinion as a bookmaker, but you set odds correlated to the percentage chance that it's *the* answer through the door in the future, not set odds *that* it's *your own* answer in a binary sense. That would be ridiculous.

Clearly you are not being asked to be ridiculous as a doctor, psychiatrist, or judge either. You don't estimate that it's *your own* answer correlated to *your own thoughts at the time*. "this gentleman *looks* a bit rapey", or "this woman *looks* a bit like she might be schizophrenic". You have to do the work. The clinical assessment.

The clinical assessment (doctor, psychiatrist), the judicial verdict (judge), the creation of odds (bookmaker), the compilation of the dictionary (er, dictionary compilers) are not your own personal opinion. Your mind comes into it, but 96% of people are aware of the job they are being asked to do. Therefore as a psychiatrist I know it's my own clinical assessment, not somebody else's. However I also know that any (my, or somebody else's) clinical assessment is linked to objective clinical reality, not my own thoughts at the time of how it affects me personally. It would be patently ridiculous if someone I had diagnosed as having paranoid delusions of thinking they knew Sir Alan Sugar from the television personally (for example) was externally proven to be a good friend of Sugar's and I said "the idea he's a friend of Sugar, that's the most upsetting wrong answer ever" meaning *to my own* answer at the time prior. That's not how it works. I don't decide the answer based on what I want or had already decided. I have to be flexible to the objective truth. I can't bend the objective truth to fit my own personal truth of the past. Well certainly not as much as that! To say that someone who clearly knows Sugar and can prove it, is still "the most upsetting person ever" because they upset *me* in terms of *my own* answer!

7.5 The Hybrid Argument

The Hybrid Argument, and how sociopaths are diagnosed

The hybrid argument can be explained using real life neutral settings such as Wembley Stadium, or any set, for example again our Knightmare set from ITV Plc in 1990.

The difference between the Libertarian approach and the sociopathic approach is that 96% of people use *the* answer, the answer through the chamber straight ahead you can't see, to correlate objective reality to. 4% of people use *their own* answer in the here and now, as a binary decision.

This creates either:

Libertarian approach: an A/B/C left door centre door right door matrix where 96% of people use A/B/C systems and B the answer through the central door (mixed bag of marshmallows) is objective, and 4% of people use A/C systems and have psychopathic disorders.

Sociopathic approach; 100% of people use left door right door systems which they label as an A/B decision (but would be A/C from our perspective and they are missing a middle zone). The sociopath is a vanilla strip short of a full Neapolitan from an external perspective, or a navigator short of a full rallying team compliment from an internal one. (only A if they think left) (only C if they think right) (but should be A/B A is subjective B is objective if the y think left) (and should be C/B C is subjective B is objective if they think right). They are driving around on dodgems, but they think everybody else also is driving around on dodgems only thinking once. Never the second thinking of how it affects the external reality.

The hybrid approach, which often pops up in mental health facilities across our country, is to use the sociopathic/psychopathically disordered system of left door or right door only (depending on which one you think it is) but in situations where everybody else would agree with you personally subjectively, say "it's objective because everybody else thinks it as well". Typically the sociopath would figuratively after the event that's gone wrong supply a big envelop to the psychiatrist stuffed with personal opinions of many other people showing any amount of people agreed with them personally.

Now this can be true, and often is, that everybody else agreed with them personally in the room (Knightmare example) that it was the more likely answer of the 2 in a binary sense (left or right) correlated to the *personal thoughts* of not just the 1 individual, but every (or most) other individuals, individually.

So "how can that not be objective?" the sociopath asks.

To which the answer is they have only done 50% of the work, being generous. And it's not the important 50%! Their mind is disordered and they think all they need to do is provide their own answer in the here and now. As if it's a short cut.

Correlating to the external answer in the room ahead of us we can't see, and hence can't know for sure, the claim could be totally ridiculous that it's 100% certain to be *that* answer, and usually is. Usually is, since there's a big difference between over 50% that you think it and hence "100% the answer" to it actually being 100% the answer, objectively.

For example, if we use football teams. And now give you 3 figures to objectively estimate

1: the apantasi answer (linked to the answer through the chamber straight ahead we can't see)

2: the dou apantasi answer, to you personally. (the personal answer in the room, but only your own personal answer)

3: the dou apantasi answer, of everyone (the personal answer in the room, the percentage of people who agree that it's their own answer that it's more likely)

I'm still visualising the room in Knightmare, you could if you prefer now visualise Wembley Stadium.

Aston Villa (left door team) vs Brighton (right door team)

1) chance Aston Villa win 52% objectively (the answer in the future)

2) chance Aston Villa win 100% objectively (THAT it's my answer in the here and now of the 2)

3) percentage of people who think Aston Villa will win 60%

Aston Villa (left door team) vs West Ham (right door team)

1) chance Aston Villa win 61% objectively (the answer in the future)

2) chance Aston Villa win 100% objectively (THAT it's my own answer in the here and now)

3) percentage of people who think Aston Villa will win 80%

Aston Villa (left door team) vs Birmingham City (right door team)

1) chance Aston Villa will win 75% objectively (the answer in the future)

2) chance Aston Villa win 100% objectively (THAT it's my own answer in the here and now)

3) percentage of people who think Aston Villa will win 97%

Aston Villa (left door team) vs Solihull Moors (right door team)

1) chance Aston Villa will win 97% objectively (the answer in the future)

2) chance Aston Villa will win 100% objectively (THAT it's my own answer in the here and now)

3) percentage of people who think Aston Villa will win is what they think (of the 2), 99.95%

Now this isn't incredibly bad so far in terms of the likelihood of practical negative consequences, but this next one should highlight the problem with it of how it specifically makes the worst errors possible against the worst people to make them against:

Aston Villa (left door team) vs team featuring Solihull U8s guy from 2005 looking 20 years into the future where the result of the game will be, who are theoretically certain to win in B, but the Manager of Aston Villa "doesn't know that" using her A/C system because she doesn't know how to look at it correlated to B because of her psychopathic disorder.

1) chance Aston Villa will win meaning THE answer - 99.25% objectively

2) chance Aston Villa will win meaning HER OWN answer THAT it WAS her own answer - 100% objectively

3) percentage of people who think Aston Villa will win in 2003 is what they think in the here and now in 2003 against Solihull U-8s guy 100% objectively, or 99.99999999999999% of people. All but bizarre mavericks.

So if you think objective means the answer dou apantisi, because you have a psychopathic disorder, you can also prove that 100% of people are 100% objectively certain it's 'THE' answer. Precisely what both of our culprits Lady MB and TH claimed some time ago, say in 2003, that the answer is whatever they thought and infinitely ever after, as whatever they've decided is the 'objective' answer because their minds are disordered.

And just as interestingly, they would still claim it vs Jack Grealish and Manchester City in 2025 even after the answer (the outcome, the external answer) that the under 8 from Solihull was Jack Grealish and plays for Man City.

Because "objective" to the sociopath is *only their own answer in the here and now of when they took the original decision* which was in 2003 long before the present year, and could *never* be changed because of any apantasi answer. That would be like "looking for blood at the scene of the crime".

If the chief of the Secret Intelligence Service is a believer in this hybrid dou apantasi system, Mr M, and Mr M is telling us the truth that he truly thinks it, M also has a psychopathically disordered mind.

How Sociopaths are diagnosed

The easiest way to diagnose a sociopath, and some 300 million people in the world are, is to ask them to *estimate* the objective answer.

If they estimate *that* it's THEIR OWN answer, they have a psychopathic disorder. 100% of the time will be 100% that the white marshmallow or Aston Villa win is 'The objective' answer. Synonymous with *that* it's THEIR OWN answer

If they estimate *the* answer, e.g. 70% or 97% chance Aston Villa win and that it could be a pink marshmallow, they don't have a psychopathic disorder.

It couldn't be a pink marshmallow in the white bag of marshmallows (A).

7.6 Four Key Truths

Self-righteousness/ego-opinionatedness of believing a subjective opinion is right objectively if you think so.

The source of all human error!

But sociopaths do it 100% of the time! And then leverage it with whatever power is within their hands.

Saying "it doesn't make any difference"..

because "that's what everybody else is doing as well" as if 7.7 billion people in the world behave like mad Roman emperors.

On Liberty: what is Objectivity?

If you wanted better examples of how to conduct objectivity, I would propose the liberal and libertarian principles of On Liberty by John Stuart Mill.

"It is the duty of governments, and of individuals, to form the truest opinions they can; to form them carefully, and to never impose them on others unless they are quite sure they are right" p30 On Liberty.

"There is the greatest difference between presuming an opinion to be true, because, with every opportunity for contesting it, it has not been refuted, and assuming its truth for the purpose of not permitting its refutation" p30.

"Complete liberty of contradicting and disproving our opinion, is the very condition which justifies us in assuming its truth for purposes of action; and on no other terms can a being with human faculties have any other rational assurance of being right".

"on any matter not self-evident there are ninety nine people incapable of judging it for the one that is" p31 "and the capacity of the one that is only comparative".

"the most intolerant of churches, listens to a 'devil's advocate' " p31.

"If the challenge is not accepted, or the challenge fails, we are far enough from certainty still, but we have done the best;.. Neglected nothing that could give truth a chance of reaching us".

"Strange it is, that men should admit the validity of the arguments for free discussion, but object to their being 'pushed to an extreme'; not seeing that unless the reasons are good

for an extreme case, they are not good for any case".

"Strange that they should imagine they are not assuming infallibility, when they acknowledge that there should be free discussion on all subjects which can possibly be *doubtful*, but think that some particular principle or doctrine should be forbidden to be questioned because it is so certain because *they* are certain that it is certain" p34.

"There are, it is alleged, certain beliefs, so useful to wellbeing, that it is the duty of governments to uphold those beliefs" p34. "There is the same need of that infallible judge

of noxious opinions, unless the opinion condemned has full opportunity to defend itself".

"There can be no fair discussion on the question of usefulness when an argument so vital may be employed on one side but not the other" p34.

p36 "it is not the feeling sure of a doctrine which I call an assumption of infallibility. It is the undertaking to decide that question *for others*".

"However positive one's own persuasion may be, not only of the falsity but the pernicious consequences, - not only of the pernicious consequences but the immorality and impiety

of an opinion, yet in pursuance of the private judgement though backed by the public judgement of his peers or country, he prevents the opinion from being heard in the defence,

he assumes infallibility".

If our detractors think they have read this, they have badly misunderstood it. The defence is not being allowed to exist, let alone be heard! In stone cold print Mill is saying they assume infallibility, and no matter how sure someone is, it is only their opinion, and the defence should be allowed to flourish so that we can examine 50% of the argument!

"The persecutors of Christianity were in the right that persecution is an ordeal the truth ought to pass, legal penalties being in the end, powerless against truth. This is a form

of the argument for religious intolerance, sufficiently remarkable not to be passed without notice." p41

"A theory which maintains that the truth may justifiably be persecuted, because the persecution can possibly do it no harm" p42.

Yet another form and variant of *infallibility*.

"History teems with instances of truth put back by persecution. If not suppressed for ever, it may be thrown back for centuries" p43.

"Persecution has always succeeded, save where the heretics were too strong a party to be effectively persecuted".

p70 "it is a great measure the opposition of the other that keeps each within the limits of reason and sanity".

The Four Key Truths p77

“First, if any opinion is condemned to silence, that opinion may, for aught we know, be true. To deny this is to assume our own infallibility.

Secondly, though the silenced opinion be an error, it may, and very commonly does, contain a portion of truth; and since the general or prevailing opinion on any subject is rarely or never the whole truth, it is only by the collision of adverse opinions that the remainder of the truth has any chance of being supplied.

Thirdly, even if the received opinion be not only true but the whole truth; unless it is suffered to be, and actually is, vigorously and earnestly contested, it will, by most of those who receive it, be held in the manner of a prejudice, with little comprehension or feeling of its rational grounds.

And not only this, but fourthly the meaning of the condition itself will be in danger of being lost, or enfeebled, and deprived of its vital effect on the character and conduct the dogma becoming a mere formal profession, inefficacious for good but cumbering the ground and preventing the growth of any real and heartfelt conviction, from reason or personal experience."

7.7. Does it Work?

If we take the two arguments, both have 1) a systemic belief and 2) a policy within that belief.

The non-consent basis holds that the person with the power takes the decision but is not held to account for any error from that decision externally as "they didn't know" subjectively, and life is on a left door right door system basis for decisions, having previously got there from the previous left door right door decisions.

The consent left centre right matrix holds that the person with the opinion is internally responsible for their own decisions and the consequence of those actions externally. And thus the aim of the game is not to take the brash subjective decisions that cause all sorts of suffering, with everyone else causing all sort of suffering and "may the strongest survive" but to gatekeep against it. You are the guardian or gatekeeper.

Modern day approaches might be Japan's philosophy towards its powerful neighbour China, where they are not so much protecting themselves by a military arms race with the country that is slated to be 18 times more economically powerful than Britain by 2045, but trying to defend conceptually by alliance.

I am not so much advocating that that is an ideal approach to dealing with military defence, of having to deal with someone more powerful than yourselves by 18 times and throwing your hands up in the air, but that military defence is an extreme example, and the level of defence an individual requires is rare to be violence, and that we do, in fact, take the Japanese civil approach, some 99.99% of us in our day to day lives as individuals without realising it. Very few people wear bullet proof vests, most rely on the conceptual protection of the Police and Judiciary rather than the literal protection of an arms race with each individual one encounters in London.

**

"without them, human life would become a stagnant pool" p94.

"There is only too great a tendency for the best beliefs and practises to degenerate into the mechanical" and "unless there were a succession of persons

whose ever recurring originality prevents the smallest shock making that entire empire die out" p94.

"Persons of genius are, it is true, rare, but in order to have them it is necessary to preserve the soil in which they grow. Genius can only breathe freely in an

atmosphere of freedom".

"Persons of genius are more individual than other people" p94.

"I insist thus on genius, most people are, in reality, indifferent to it" p95.

Only The Lady and Mr H, however, misunderstand that that concept can only apply to themselves and not someone else that they don't agree with. "This genius isn't a genius because *I* don't understand his genius!" I see it now, that Founder of Google, now I see Google, but *I* didn't share his genius to create it and thus it is not my fault I condemned him as worst of all worst people because *I* didn't know and objectivity is subjective to me.

"What is it that has hitherto preserved Europe? What has made the European family of nations improving, instead of a stationary proportion of mankind? Their remarkable diversity of character and cultures. Individuals, classes, nations, have been extremely unlike one another, but they have struck out a great variety of paths, each leading to something

valuable" p105.

compared to "the Chinese ideal of making everyone alike" p106.

"If the claims of individuality are ever to be asserted, the time is now, while much is wanting in the enforced assimilation" p108.

"Mankind has become speedily incapable to conceive diversity, when they have been for some time unaccustomed to seeing it" p108.

I think Mill is saying that the homogenous Chinese compared to the Europe he was describing as successful and tolerant and progressive is a bad thing for humanity that quickly leads to a loss in Liberty. Not a great blueprint formula as our racial puritists seem to think in condemning anyone that is not 100% White from birth for lack of "Chinese purity" oh I meant "racial purity".

Yet Security Services argue for the insistence of the right to make us completely alike like the Chinese Communist, in spite of what makes the entire continent of Europe intellectually successful!

7.8 I didn't know, I didn't know he could be

Algebraic (needs to be certain absolutely) vs linguistic self-righteous (needs to be certain inside my own head and if you disagree with me you are wrong and arrogant) vs *linguistic* sociopathic (needs to be certain inside my head, and if you disagree with me you are wrong and arrogant and need to be punished)

The algebraic thinker cannot punish somebody for being wrong, only hold their ground in defence against error. As error is acknowledged as certain and not punished, only the arrogance of accusing someone else of arrogance and punishing them can itself be punished!

So the algebraic thinker would never be aggressive or non-consensual, only strong in defence.

They would only ever think they were right, if it were certain by absolute algebraic terms that it *was* right.

There is a difference between saying "I didn't know" after:

1. "I didn't know", non-consensual, subjective, aggressive.

and

2. "I didn't know" consensual, objective, non-aggressive.

e.g.

1 I didn't know not to insist that I was 3000 times better than a Russian chess player at chess who I didn't know was a chess player, say he was "the most upsetting of all upsetting people" because of what he thought of himself, insisting I myself am correct, put him in a fate worse than jail for 30 years linking it and leveraging to a nuclear missile against Russia if they don't accept my subjective analysis.

2. I didn't know my kettle needed descaling.

7.9 100% of arguments

Algebraic approach would never punish by *positive* intervention somebody else for being wrong in their non-aggressive thoughts or actions, as this is an expected part of algebraic reality. It is important to note that *every argument in the world is still had unless you forgive ego in others, as well as control your own ego.*

Every argument only needs 1 person to be wrong. So unless you yourself forgive someone else who has something intellectually wrong you can put it in x terms again.

Person A wrong (subjective objective) X wrong (objective subjective) meets Person B right = argument

Person B wrong (subjective objective) X wrong (objective subjective) meets Person A right = argument

Person A or B is wrong, Person A or B right but lets it go = no argument

so it is always the person who is *right* that decides not to argue, because you cannot help the ego of someone that is wrong existing.

The onus is on you to be sensible and *help* that person and the world, not *harm*/be as *difficult* to that person as possible. You are also wrong *yourselves* sometimes you know!

Every argument in the world is caused as x when you don't let off somebody else's ego.

And every argument in the world could be saved as x when you *do* let off somebody else's ego.

7.10 What to do about it

One could say that the 2 biggest theoretical reasons for existing in general terms:

1. Control your own ego.

2. Help other people control their own egos by *not* responding. Meditation helps.

You could never cure ego by punishing someone for their ego. They don't know they're wrong. *You* are the only one that knows that you are right, thus *you* have the moral obligation.

But you've read this book and they haven't, so you since you know it already, or now know it from reading this sentence, also have the obligation to control your own ego as well as not reacting to theirs. Whereas they don't. That may seem unfair, but I don't make up the rules!

The brotherhood of man will be better for your efforts if you play by these algebraic rules of letting other people off, as well as trying to improve yourself. Not easy to do. Not necessarily "fair" either unless we try to be as enlightened as possible and consider that other people's egos are really the business of themselves, so if they irritate us expressing their ego that's really a matter for them. They will be punished for using ego, it logically follows using our mathematical theory, they should also be keeping themselves trapped in the simulation to be asked again to learn or repeat the error. It's why other people are here too: either to be considered positively as a learning experience to figure out not to do it, or negatively they are also trapped here by their own egos too.

That's it then. The purpose of all this is that there is probably some sort of choice to be made, whether we want to act in the right way and be rewarded possibly later, or to act in the wrong way and vulture everyone else and receive material benefits. "To decide". It's amazing that we've lasted this many thousands of years without anyone really being able to remove the tramlines of this decision between good and evil, right and wrong, or intellectual truth vs ego-aggrandisement which can always be decided by the self.

CHAPTER 8 - Case Studies

One happened in real life

One is fictional

8.1 The Government of Britain vs Mixed Race Royal

8.2 Someone must prove they are the most intelligent person in Britain/the World or Best Chess Player in the World if they upset the "most intelligent people" running our country or our country's chess?

8.1 The Government of Britain vs Mixed Race Royal

Real Example - Mixed Race royal

There have historically been examples of white supremacist thinking including slavery, genocide of various tribes and races, the Partition of India, and secret organisations such as the Knights Templar with mottos "racial purity is the most important thing". Can we really be sure that all traits of racism in the world have been eliminated, or perhaps have some of these organisations gone undercover in the modern age?

When a mixed race woman joined the Royal Family of Britain, the couple were ostracised and relieved of many duties, as if the powers that be in Britain felt this modernising touch for the Royal Family was something that needed to happen, but they could tolerate it so little that as soon as it did, negotiations terminated for any future active involvement in royal duties.

Some of the comments included UKIP activists writing a "Twitter" post on how "our Royal Family will be tainted by Black seed". While these offerings were lampooned by liberal (female White woman, and Asian male) comedians on the BBC, Britain is the least fair country in the Western World, you are 55 x less likely to go to the best universities from a poor background, it's the only country in the developed world that has gone backwards in the great global age of meritocracy, and 1 in 7 white men in London are a member of the Freemasons, with association with the Knights Templar that said "racial purity is the most important thing".

Introduction/Recap to theory:

1) It's algebraically impossible to be the other person's ego if you prejudice the initial variable and someone has upset you the most in your life. Because *somebody else's* ego cannot by *reflection*, switch on your own ego, *more* than any *real ego*.

When we think we are right, but we are not, the effect on us of our own real ego is worse than someone else simply being wrong. It's the *perception* of the false negative when someone *is* right but you didn't *know* they were, like the fable of the dog looking into the reflection of the other dog stealing his bone. The dog that annoys us the most is ourselves!

2) For someone to be the most upsetting wrong person, or an idea to be the most upsetting wrong idea, they would have to be the *worst* of the *worst*, but also more upsetting than any real ego of our own, if they're not right.

Even if this *was* the unlikely possibility in the first place that the person was the most upsetting wrong person...

they'd also *have* to be the worst of everyone *right and* wrong. The idea that non-White people could be the "the worst of all wrong things and worst of all wrong AND *right* that we didn't know were *right* things", seriously?

There follows examples of evidence to counter the idea that "racial purity" is superior, using purely the algebraic approach.

Scenario: Right-wing extremists in the British Government have established that a mixed race royal has married into the Royal Family, and think, using linguistic logic, that this is "sullying the Royal Family against their own will". No doubt with phrases such as "national purity" being bandied about that this would be a popular viewpoint, with 2 billion members of the world approximately being connected to a racial purity cult, The Royal Arch of Freemasons, that toast "for purity".

1. Problem in Question of Definitions

the phrase "racial purity" is a bias. The non-biased phrase would contain no positive word, so "100% White" is accurate, or if you want to be wordier "100% of any race".

But to say "racially pure" obviously contains the conclusion to the question within it, thus should not be taken seriously.

2. Problem in Question of Misclassification. Is a 99.9% recurring White person a White person, or a mixed race person? Or something else, specifically new if she can't be

both?

Looking at dogs: a Newfoundland dog that breeds with another type of dog could never become a Labrador?

Does a European that travels on a boat across a sea west become 1) a " 'mongrel' European" or 2) an American.

Is an American a mongrel European? There is a more accurate delineation within that description that they are an American.

Certainly I would not expect an American to be *denied* the right to call themselves an American by Europeans calling rank on an intellectual question that pertains to the other.

But 100% White people are more powerful than 99.9% White people, whereas Americans are *more* powerful than the European countries they came from. Therefore it seems more likely that the less powerful mixed race person is going to be oppressed by the more powerful White people running Britain, than the American is going to be (going to be *able* to be) oppressed by the British person. Well, at least these days.

3. Accusation is barking mad combination of false definition and classification and bias

"The birth of a mixed race person, even one 0.1%, is worse than murdering 7 people where they "sully" the Royal Family against their own will". These phrases are used by Mr H, and consistently by the likes of UKIP members "taint our Royal Family with her foreign seed" was picked up by BBC1 The Mash Report.

Several problems with this:

- Misclassification with potentially pejorative phrases such as "mulatto" may not be an accurate reflection, simply because racial puritists have decided they want to classify someone in the potentially pejorative fashion.

- The intellectually racist claim ends up on the wrong side of its own "value paradigm" later explained by saying mixed race people or slightly mixed race people, or Jewish people (the most internationalist diaspora) are inferior because they don't like them. That they don't like slightly mixed race and/or Jewish people does not prevent mixed background Jews Bobby Fischer and Garry Kasparov annoying them at chess because they are *better* than them, or that the same claim that Jews have 400% more capital than a White person on average may also be viewed positively as a success, not evidence of conspiracy.

"Sully" the Royal Family against their own will.

- You cannot 'self-sully'

- you are either part of the Royal Family or not. Logically you cannot change 'perspective' mid-way through.

The claim says that you as part of the Royal Family 'sully' the Royal Family where you are then not part of it as the act of consensual breeding as part of the Royal Family could in no way not be part of it.

You could say "you sullied the opinion of the King, against her own will" but that wouldn't make particular sense. It makes more sense than the claim which changed perspectives mid-way through.

Nonetheless, let's refute it!

1. HORIZONTAL ERROR

Every other subject in history, the algebraic structure of this argument would end up being proven wrong at a later date.

Immediately, Copernicus is a good example. Does the earth revolve around the Sun, or the Sun around the Earth as "everybody else" thought?

That Copernicus said Earth around Sun.

if the Sun really *did* revolve around the earth there would be no algebraic *friction*. Thus the *wrong* answer would actually *not* annoy.

The *one* answer that later got proven *right... annoys* with *friction*.. so walk the plank.

And every other answer in world history that got proven right later also follows the same algebraic structure. It was opposed with *friction*. Not opposed with '*non-friction*.'

When anything was opposed with non-friction, i.e. not algebraic *velocity*, the thing doesn't get proven right later. It was just wrong.

So can something contain algebraic velocity and be right? No.

Can it not be so lame and so obnoxious that it becomes right? No.

There are 2 key elements to it:

a) that the subject is inferior.

b) that that inferiority upsets them the most in their lives.

But they say that the combination of those 2 things is not algebraic velocity (impossible! in the algebraic approach)

a) If a subject is inferior, and true in the world, it produces a reaction of sympathy.

If a subject is *inferior but untrue*, e.g. I make it up by saying someone has a "Jif Lemon for a Head" it produces a consistent reaction of laughter/smiling/non-caring.

b) if a subject is a secret superiority, it 100% produces a reaction of intellectual annoyance.

This algebraically suggests with certainty, that since this is the only reaction that race mixing could produce if it were an unknown marginal good for the species, as well as the only reaction produced in every wrong thing in every other subject in history (horizontal subject comparison) that the thing that is producing such velocity as “it’s worse than murdering 7 people” may, yet again, like Copernicus, like every other daft Copernicus-style inquisition in history, be another right answer of a novelty for the world.

2. VERTICAL ERROR

100% of other weaknesses in the vertical category of the subject itself - genetic weakness, produce a reaction of sympathy.

e.g.

- spina bifida

- cleft palate

- heterochromia (debatably a weakness)

To be mixed race and that be "inferior" but to produce a reaction of intellectual annoyance, would be *inconsistent* with any other genetic weakness.

Which are *all* consistent with *each other*.

So “race mixing is inferior” argument is:

1) consistent with every *wrong* accusation in every other subject.

2) inconsistent with every *right* answer of genetic weakness in this subject (unless you didn’t like that shade of chocolate? I assume genetic inferiority of the non-White is being claimed).

b) which are all consistent with each other!

3. Other Examples

In trying to find other examples where human beings would condemn other human beings from birth, with velocity, the closest I could get are related to war. For example in areas with ongoing tension such as the Middle East, it is unfortunate that some people are brought up on both sides with an ingrained attitude of hatred and resentment. Any longstanding dispute also Armenia vs Azerbaijan produce hostility. However the hostility is not a velocitous hostility of insisting that the other side are the most upsetting wrong people ever, generally, to every other people and country in the world. More of a personal dislike, based on historical reason whether valid, partially valid or invalid, is understandable. We can see the claims of such countries against each other is understood in the minds of each side that the other is more the most upsetting wrong tribe to *us*, as opposed to every other individual of every other nation in the world!

It's truly remarkable therefore, that Britain can make such claims against mixed race people, or 99.9% White people that the children of war would never dream of making, and think they could be objectively correct in such ridiculous statements.

In Britain's case, the intellectual annoyance is *real*. It's their ego. And you can find no ego in anywhere else across the world in which they would *genuinely* condemn someone being born as worse than killing 7 people and *mean* it with only the *flimsiest* of excuses to mitigate of "it's because it affected the Royal Family".

That's flimsy intellectually, as affecting the Royal Family makes no difference to the intellectual issue. Same as whether it affected the US President, God or space aliens.

"You sullied God against his own will by being Welsh" (that's me, by the way) doesn't make that sentence *make* sense, if it didn't make sense intellectually in the *first* place.

Argument 4: Scientifically in nature the thing that takes the longest to develop is always or usually the best.

Mixed race people (for example those that are 99.9% recurring percent White) would thus, taking the longest to develop, be more likely to be more intelligent than less intelligent than White people by the evidence of nature all around us.

100% White people are arguing that "mixed race" people are inferior, but 100% White people could be seen as the caterpillar, that eventually becomes the butterfly, which was really the aim in the first place. I am making a positive argument for being mixed race, I know it is unusual, but we barely hear any discussion of any potential positives such as in preventing genetic disease, but we have surely heard in history enough the counter-argument!

Will a new breed of dog ever be as intelligent as the existing breeds of dogs that already exist? And were all breeds of the past more intelligent than new breeds?

Argument 5: IMMIGRATION vs FRENCH

Let us consider if White people who claim mixed race people are inferior say "we are protecting our race" or doing something innocent the 2 side by side comparisons.

1) How it affects other races that are not your own race (is your attitude the same?)

2) What about things that change the demographic in the same way as having 99.99% r White people generally, but in another way, that does not include the *process*.

They do not give a flying toss about this desire to protect "racial purity" if it is the racial purity of another race where it would not affect *them*.

I.e. yes the British racial puritists would *claim* that Japanese purity is important to them too, if only as defensive measure against the "mongrels".

But would they *care* about *Japanese on the Moon,* 25 million years away for example. If only Japanese as a nation were on the Moon?

If the British were on the Moon, their own daughters, I put it to you algebraically that they would still care. If the Japanese were on the Moon they would let the Japanese get race mixed by space aliens without giving it two intellectual hoots, as long as it didn't interfere with the English on the Earth.

Argument 5b: Immigration from Africa that is not Black

If it is not velocity, immigration would have the same effect on the "intellectual annoyance" of a 100% White person, since these people coming in are *also* "sullying" the purity (reminder: I'm quoting the racial puritists, and opposing their argument).

However, White immigrants from Africa don't do this, so it is not being *African* that does it either.

Argument 5c: Immigration from Black Africa.

If it is not velocity, Black immigrants would have the same effect on the "intellectual annoyance" of a 100% White person coming into Britain, but while this gets closer to upsetting the intellectual racist, it does not "intellectually annoy". Or it does not

50% as much.

So it is not even being *Black* (like I myself am part Black) that is the issue.

Argument 5d: if the world was close to being racially pure, it is most likely to be 100% Chinese and/or Indian and racially pure after a partial genocide of other races including the White race.

If the world was 100% pure and 99% - 100% Chinese, but Britain was kept 100% White and made up a similar amount of the world, would the White racial puritists, be happy with the racial purity achieved?

What if we throw in just Madagascar (a former plan of theirs for non-Whites), or perhaps Northern Ireland, for White people?

So a world that changed from racially impure to 100% racially pure did *not* actually have the effect in you of pleasing you incredibly , or even at all, even though it changed nothing of the thing you claimed to want negatively, and did everything to make happen what you insist you want (racial purity, and Britain being 100% racially pure).

Argument 5e/conclusion:

So you see it is the *process* of race mixing, because it produces the 99.9% recurring people of your own race. I.e. 99.9% recurring White people for the White race.

And only when the *process* that produces the *algebraic superiority* by inverse deduction (from velocity) as well as all the other evidence, you see this velocity.

It cannot be "not velocity because you are not wrong". That *is* velocity.

This is an intellectual prejudice on an unknown outcome! This is absurd! Prejudice of the only variable before the result saying "we don't need to not prejudice it as we know what the result will be".

Wrong on 2 counts:

1) you don't know what the result will be. In absolute terms.

2) only someone who is going to be wrong ever prejudices a result (seeing as how anyone that was going to be right anyway would let the result happen). This is called *velocity*. "I don't need to play this chess game against Weinerstein/Kasparov because I know I am better than him".

Why are you only better than *him* where you don't need to play him to prove it?

"Because he's the most upsetting" "worse than murdering 7 people" person.

But that might be - and theoretically *is* - because they're better than you.

"We know we're better". No you don't. You know *you think* it.

Argument 6: Does God/Aliens know more than us?

Various races exist and may have been put here by God/aliens.

It is the height of arrogance and inability to think objectively in an *objective* sense, to only look at it from *our own* sense of what our own position and say that we cannot work out the main benefit/purpose, and thus that it could not exist. "We absolutely know".

Anyone that doesn't understand that they can't decide this argument for themselves against the greatest unknowns, is self-righteous and possibly sociopathic, but a lot of people are. Like "would you listen to a message from God, or the overseer, if you knew it was them, but you didn't agree with it" sort of in generalised form. It seems that something a lot more powerful than us have created various races, and some form of intermingling was an absolute inevitability. So while I cannot be certain that this was deliberate, I can be reasonably certain that it is probable that some power greater than us didn't have such a great problem with it.

Then again I do *know* a lot of people who wouldn't listen to that message from God *either* such as Mr H, if it stopped them doing what they wanted, so trying to make a point by comparing it to them may fall on deaf ears...

I don't see too much else sent by God/aliens/mathematicians being argued with as something that is "worse than killing 7 people" except for other things that we ended up being wrong about such as killing Socrates.

But that's the algebraic approach. And how can you get linguistic thinkers to accept algebraic reality?

It's like trying to explain "there is no scientific evidence for the existence of astrology" vs "there is no astrological evidence for the existence of science" where they are the one arguing that the potentially bullshit interpretative one is the one that's right (words and language) and the pure tautologous reality of scientific mathematical reality is the one that is wrong but they in turn think it is the other way round that we can't claim mathematical certainty might be correct just because it's theoretically certain, when *they know* that nobody agrees with it in the linguistic mode at Masonic dinners.

When I was at a Masonic dinner, at Ladies night, I heard the conversation "I'm in charge of the Lodge newspaper, I like to think of it a bit like the Daily Mail" by the Master named Mike, and his daughter replied "yes except it is more right-wing and less hospitable to immigrants!"

Argument 7: Any other Speculation

Could it be a reason not covered?

If we look at the game of chess: chess makes humans more intelligent than humans trying to teach humans how to be intelligent. Thus some people speculate (including the long-time President of FIDE) in all seriousness that it may or must have been created by aliens. Interesting stuff. Hard to really argue against the possibility. Maybe chess players are actually more intelligent than everyone else too, by that logic, but the world is geared more socially to other people?

To go a little further on the chess was created by aliens idea, which was necessary to mention for the following reason:

If chess *was* created by aliens, those aliens are more intelligent than us and are trying to tell us things, and they are much more algebro-electromagnetic (telepathic?) than the human mind.

I notice using the game of chess that the aim of chess, especially in the early stage, the opening, is to build a preponderance of domination on a particular colour complex: these are called "light square openings" and "dark square openings" but have access to *both*.

It's a bit speculative, but I wonder if they were trying to tell us something. Maybe the aim of chess is for your chess pieces to *avoid* becoming racially pure but can be as *'racially pure'* as the like, without actually becoming 100% racially pure one colour only and hence draughtsmen?

I won't go any further with that but to tell you that I just gave you 1 idea, but there are 250,000 reasons playing on both colours could be better for your chess pieces than just playing on just the white squares, just none of them are provable the same as the other way it can't be proven absolutely. But the other way round, there are billions of people, in high ranking positions who believe that they have it proven. They don't, and *we* mixed race people have to defend ourselves from people insisting it is.

We are on the defence, as there are no "mixed race superiority" organisations terrifying White people with 2 billion members..

it is more a matter of perspective to understand that being on the defensive with no potential attack and having to defend perfectly and then hearing "the Black man's got a chip on his shoulder" if anything is ever said. Why shouldn't we be at least equal?

There are no definite proofs either way. But the algebra is 100% certain that the racial purity cult is wrong. And John Stuart Mill said that we only understand 50% of our own argument by allowing our opponents' ideas to exist.

The prejudice of insisting "racial purity" is better when it is a variable, with velocity, and then that the velocity is not velocity (which prejudices the only variable by insisting its right) because "we're not wrong" is all great fun (nonsensical) stuff. But realistically you have to stop claiming it, and probably go looking for an answer *the other* side of the argument to get anywhere now with this, now we know it's 100% certain algebraically, once we understand the algebraic approach, it is the *other* answer. To claim something with prejudice of the only variable, that the weak 'inferiority' is giving you a tautologous emotional and algebraic reaction of a superiority you don't know is a superiority, where all other inferiorities produce a reaction of sympathy? We know they are wrong using the algebraic approach.

Argument 8: Sullying the Dog Royal Family. (Perspective)

I don't have to prove that 99.9% recurring White people are superior to 100% White people, because I am not claiming it in aggression.

The only reason I know it's 100% certain algebraically, is because 100% White people claim with 100% certainty that they are superior based on 3 things at the same time:

1. They claim they are superior in a way that prejudices intellectually the only variable. And we absolutely do not know the answer in absolute terms (that also prevents me from giving it of course, but at least I'm not daft enough to say I have it by prejudice).

Thus we find the British Government on the wrong end of their own sociopathy again. Where they are making an unsubstantiated claim by final conclusion.

They say they have evidence, but the evidence could never be fully in on the difference between 99.9% recurring and 100% and until we have that it is *very* much in debate as I have shown from the previous 7 examples.

2. They claim a 99.9% recurring White mixed race person is "worse than murdering 7 people" by intellectual weakness.

That's velocity. I covered this in Chapter 5.

They say it's not velocity as it's only because it affects a member of the Royal Family. But that wouldn't influence the issue of whether it's velocity.

For example: if someone didn't like Mackintosh's Quality Street and then sullied the Royal Family by giving a definitive gene that *definitely* would make Princess Anne like them (or indeed the other way round to *not* like them if the other person did) then *that* ***does*** *affect* the Royal Family, but it would not be claimed to be "worse than murdering 7 people", as there would be no velocity.

Neither does being born with a Jif Lemon for a head get the aggressive velocity of being worse than murdering 7 people. Which definitely *is* worse than being Black, White or mixed race. Even if King Charles was sullied in such a way, by a Jif Lemon sullying him by squirting itself and asexually reproducing with him, it wouldn't produce *intellectual* annoyance, seeing as it's not *superior* to have a Jif Lemon for a head.

So you can see it is only a superior truth that's unknown that produces a reaction of intellectual annoyance. You can see that no inferiority does it. You can see that nothing *false* does it.

Having a Jif Lemon for a head, is definitely worse than being 0.1% mixed race.

Thus if you wouldn't claim that a Jif Lemon head genuinely could have the effect on you of intellectually annoying you (that *murderous emotional* feeling) but someone being slightly mixed race does, but *only* when it is *your own* race and *only* when it involves the *process* as opposed to a Frenchman or Nigerian walking into the country.. that's called *velocity*. It's algebraically certain the thing that is upsetting you more than anything else like that is actually... the *right* answer... and *you* have the wrong answer.

It's *your* ego. Not the ego of the princess or slightly Black man.

Remember also: it had to be one person's ego. The slightly mixed race man or the princess before they were born then?

TIMELINE OF ERROR

1. Going against 100% algebraic theory that is tautologous because it doesn't suit.

2. Impossible (non) algebraic claim that prejudices the thing intellectually but also then insists the other person is "worst of all wrong and correct answers" instead of simply right. So now *two* bogus claims. "two wrongs don't make a right".

3. Psychopathic decision/plan of action/lack of mitigation. To then teach the person a lesson for being wrong (when they are right)...

and punishing them (3rd error) for being arrogant (2nd error). Where punishment is now the worst punishment because the person was "most arrogant".

4. Putting oneself in charge of turning back on the 0.1% possibility they were wrong. Which will never be turned back because they won't 'uninterpret' it they'll *re*-interpret it.

The guy who couldn't turn off the oven at the switch when it was 100% certain to be dangerous, gets put in charge (by himself!) of turning it off at the plug.

5. Ignoring evidence with bias.

6. Disbelieving evidence with bias.

7. Refusing to turn back even on completion of error.

8 Bias of evidence.

Even in all of this in language, where someone behaves completely anti-algebraically.

We would actually get, on evidence that contradicts his wrong belief:

- I believe but it's insignificant

- I don't believe but it would have been significant if I had believed

against a mathematical certainty that was linguistically not-believed.

I.e. the theoretical certainty was 1) impossible because it was *able* by other people's consent who *also* think in language..

then 2) this *remains* impossible and the evidence - as ridiculous as it is to dismiss – *still is* dismissed by the same initial bias that existed in the theoretical question, now manifesting in the practical one.

Remember that it has to be *somebody's* ego using the algebraic approach, as *somebody* was receiving the message that the other idea is correct using the holographic mathematical universe. That's the whole point of this book. See the formula.

So it could not have been the mixed race person before they were born. That's even more impossible than it could have been the Jews in The Holocaust.

This is the point of the book, remember, to show the algebraic message of "the code" (part 2) where we are in a simulation (part 1).

This would have made no sense if the world was linguistic and physical that we could do that. But this book hypothesises that the world is holographic and mathematical and I can show you how to do it, to work out the answer to what is just about the most complex thing ever invented (the answer to life, the universe and everything) in, hopefully, a way that gives you some chance of understanding some of what I'm on about.

8.2 Someone must prove they are the most intelligent person in Britain/ the World or Best Chess Player in the World if they upset the "most intelligent people" running our country or our country's chess?

Otherwise they're "the most upsetting wrong chess player/person ever"?

Fictional Example

Case Study 2

In Case Study 1 we looked at a mixed race royal.

Now case study 2: could a mixed race person in Britain, or a non-White person, condemned by Britain's right wing, be theoretically more intelligent than them?

The case study 1 was of the highest profile mixed race person, Princess Meghan, and hostility that met her entering the Royal Family from Britain's right wing.

Let's take a hypothetical example that would meet the same hostility from the same people. If The security services got upset, the way sociopaths think, they would compare it to their own answer as '*the*' answer and declare the person who upset them "the most upsetting wrong person ever, objectively (to their OWN answer/s)" as very many "important people" in Britain were getting upset with them.

If The Security Service of Britain are wrong in Part 2 about why there is hostility to mixed race people - which you can see that they are with 100% algebraic certainty - then you can see how this may be a problem, where they might even *deliberately* project against mixed race people individually, down to the *accidental* projection against mixed race people. They are not deliberately hostile to the idea mixed race people are equal, they are accidentally hostile, because their impulses simply tell them it is what they think. However this 'accidental' hostility can cause deliberate hostility to individual mixed race people who they are upset by, in a 2 part process. Once upset by the individual mixed race person, the underlying hostility, caused accidentally but existing, comes into play.

A Director of the Security Service insisted that Sophie, a mixed race girl, upset her a child, and since with ancestry from "another country" can't be more intelligent than her even though the algebraic approach says she is because of 3 reasons described below, they don't accept it because "she'd have to be more intelligent than the head of the intelligence service and nobody thinks she is".

Linguistic approach = makes sense "on balance of probability" someone "probably" isn't more intelligent in subjective objectivity.

Algebraic approach = is a bias.

Let's imagine someone takes them seriously, and creates a hypothetical example of could somebody be more intelligent than the security services of Britain, or do they have to be "the most upsetting of all wrong people ever" if the security services got upset. The bias on the balance of probability, makes no sense logically, because we don't know if the child is - or more likely/ obviously *could* be.

In *this* case, the bias is even more prominently displayed, because there is *no way* the child *could* ever show the ability to be "more intelligent than the Director of the Security Service" at the age of 8. It is impossible by any linguistic approach measure. If it was a chess player, the age of 8 might mean something in chess, but the age of 8 as a general intellect is like age 3 at chess to give an analogy. Even if the 3 year old became World Champion, even World Champion by 20%, at the age of 3 it is impossible to predict.

HYPOTHETICAL EXAMPLE: Mixed Race Child vs Security Services.

Let's look at the 3 reasons:

1. The Lady got in charge of The Security Service and she and everybody else that counts in Western Intelligence decided that Sophie *can't* be because they don't think so.

2. Sophie is apparently specifically "unlikely" in their view. Interesting.

3. Sophie upsets them the most in their lives (some of them) and most anyone has ever annoyed them as a group generally.

Let's address point 1.

1a) Point 1 is a bias. It has no basis in absolute proof or objective reality.

1b) All intellectual arguments are inversely correlated, like a 2 horse race with 2 jockeys, as was painstakingly explained. The horse that comes in, the jockey on the *other* horse is egotistical.

That the Security Service and Secret Intelligence Service don't think so, is pure *relativism* since the thing cannot be absolutely proven.

They're right if they're right and will think they are right.

They're wrong if they're wrong and will think they are right.

"They think they're right" as a point.

It's *meaningless.*

But it's more dangerous than meaningless, since it is a bias.

I would say that the fact that they particularly think this kid is unlikely actual veers towards proof *against* The Lady as opposed to proof in their favour, since a wrong answer cannot be particularly any more non-existent than any other wrong answer, but I'm about to make that point.

So in conclusion to point 1: that a relatively good chess player, even if that relative strength is the highest possible i.e. a World Champion, thinks they are better than an unknown opponent because they are "probably better on the balance of probability". I think it says more about the imbecility of the World Champion and asks questions of how he ever became the world number one in the first place than concluding anything definitive about the opponent!

2. Sophie was as an unknown opponent apparently "especially unlikely" in the eyes of the Security Service.

Well it gets more humorous to be honest!

Remember the thing is inversely correlated! That to Garry Kasparov encountering a secret space alien who can play chess to 3000 FIDE level but is posing as a human, the space alien would seem the "least likely" *forward slash* "most upsetting" with the prejudice of not having the prejudice-less interpretation.

A state of flux never exists in the human mind that we've already largely decided either correctly or incorrectly - certainly we *have* decided if we take "the Lady bias" of insisting we are better in the first place.

Ergo this alleged neutral state in the middle of 2 binary outcomes never truly exists like the totally universal chess player that is great in all stages of the game. Usually one inclines more towards aggression or defence/ counter-aggression.

To say that our opponent is "especially unlikely", when we

2a) don't know if they are better than us

2b) we already decided that they definitely weren't!

Well, what can I do to highlight the lunacy of "the Lady bias"? Perhaps write 2 sentences that describe each possibility and you can see how the bias affects it.

So let us take Jan Timman, the chess player who was considered "the best in the West" and we'll use position A1, A2, B1 and B2 to describe Timman as the newcomer who is right and wrong, the old hand who is right and wrong. And see how these sound:

Position A1: Timman is newcomer, and less good than the old hand who is not Timman

Old Hand "I insist I am better than Timman" (without saying it)

(encounters Timman who is worse than him.)

"Jan is the dictionary definition of arrogance, since he thinks he is the best in the West, but it is truly I, Robert Fischer"

(other guy was Robert Fischer).

Conclusion: Fischer is technically correct.

Conclusion2: this doesn't happen very much. Since a World Champion would have to say it to everybody in the world. And a World Champion *could not* get *more riled up* by one bad challenger than every *real victor* (example B1 B2 below)

The failure of the challenger is inversely linked to the success of the champion, thus if it really were a terrible challenger then the correct reaction is not to get "intellectually annoyed". And if the challenger is so bad a challenger as to not be a *contender* then it is a meaningless person that could not possibly annoy you more than any real challenger!

Position A2: Timman is newcomer, and better than the old hand who is not Timman

Old Hand "I insist I am better than Timman" (without saying it).

encounters Timman who is *better* than him.

"Jan is the dictionary definition of arrogance, since he thinks he is the best in the West, but it is truly I the old stager".

Conclusion: Old stager is technically incorrect.

Conclusion2: this happens more. Because the young challenger Jan Timman that is better than the existing status quo and everyone else with *friction* hits the *ego* of the existing orthodoxy, it hits everyone, but anyone egotistical will allow the feelings of intellectual annoyance to, well, intellectually annoy them. Some of them react to their ego.

Position B1: Harry Styles from One Direction is newcomer, and worse than the old hand who is Timman

Old Hand "I insist I am better than Harry Styles from One Direction at chess!"

(encounters Harry Styles who he is better than")

"Harry Styles is the dictionary definition of arrogance, since he thinks he is the best in the West, but it is truly I, Jan Timman".

Conclusion: Jan Timman is technically correct.

Conclusion 2: this doesn't happen. The same as it didn't happen the other way round where the x that is Timman was a or b, x is still

x.

Jan Timman is better at chess than Harry Styles, Harry Styles *if* he thinks he's better than Timman makes no difference to Timman.

Our Lady is trying to argue "but *you think* you're *better, that's what counts*". It makes no difference if they think they're better unless they *are* better.

Thinking you are better and *not* being better does not cause "most annoyance in world". It's inversely correlated. You should have looked forward to their spectacular failure, but you shouldn't have been that upset by them in the first place.

But you were both: *upset*, and they didn't *fail*.

Position B2: 3 year old or 8 year old Magnus Carlsen is newcomer, and better than old hand who is Jan Timman

Old Hand: "I insist I am better at chess than Magnus Carlsen"

"Magnus Carlsen is the dictionary definition of arrogance, since he thinks he is the best in the West, but it is truly I, Jan Timman".

Conclusion: Jan Timman is technically incorrect.

Conclusion 2: this is, if Jan Timman were ever to come unstuck, where Jan Timman would come unstuck.

Conclusion 3: but Jan Timman would not come unstuck here because Jan Timman actually never claimed this.

It is The Lady that claimed this in first place!

All I am doing by spelling it out, is pointing out how The Lady or anyone else that ever claimed this, is doomed to failure. It's only a matter of time.

If Carlsen claims it as World Champion *now - still* doomed to failure 100% certainty. He'll eventually claim it against someone that is *eventually* better than him. Be it when he's 40 years old, 50 years old, 60 years old or however old it takes him to finally be less good than some challenger simply if only because the new challenger is more in his prime by whatever ratio is necessary to dethrone best player of all time (I can keep going: 70 years old, 80, 90, 100 etc. it *will* happen!)

This "especially unlikely" business is nonsense. It's inversely correlated.

The Red ball of the casino that's "especially unlikely" does not make the black ball "especially likely" in 2 things that are inversely correlated to each other like a chess match.

It's simply a failure to understand relativism *objectively* objectively, which I described with the roulette example (above) and the 3 mathematicians example.

To think that somebody is especially unlikely, when you don't know the answer, with prejudice, is *velocity*.

Sophie, who they said was especially unlikely, is:

1 from family said to be "eighth wonder of the world" how intelligent they were

2 daughter of "world's best software programmer"

3 grandson of MI6 Officer on other side of family

4 high enough IQ to join Mensa high IQ society despite being young kid, prompting creation of "Junior Mensa".

5 frequent number 1's in music industry starting approx. age 9

6 founder of Britain's most successful retailer per square foot aged 11 strategically

7 wrote half of the music for Radiohead.

The Security Service witnessed all of the above. Did they change their mind?

They would say:

1, 2, 3: "intellect is 100% entirely genetic based, and she doesn't have the genetics

4: "IQ tests don't prove anything"

5 "that was a joke, and then mostly other people (2 Unlimited)

6. The retailer was a fluke, you were just copying Virgin Megastores".

7. They didn't bother to listen to Radiohead album saying it was "insignificant"

So these were all considered Insignificant (2) with the same bias as interpreted the fact in the first place (1) as "definitely couldn't be more intelligent than us despite theory saying so must be "least talented of all talentless people" or "most wrong

of all wrong people" which was only other option!!

Then if we head into adulthood to the last time the security services bothered looking at it at all, one grey area:

- became Europe's largest single sales artist with pseudonym "aged 13-14 creating best-selling piano music of 21st century.

They knew Sophie had something to do with this, but not as much as she did.

And so is half-insignificant (the believable 50% of it).

And half-unbelievable (the significant part of it).

Then since aged of 14, but now Sophie is in a psychological experiment where the security services are getting their revenge for her upsetting them

- created idea 400% more successful than Britain's leading company aged 14-15

- made £1.8bn profit in Venture Capital aged 18-20

- took the government of Italy aged 21-22

- created system of democracy for Russian Federation.

This would not be believed (3). With the same bias set as (2) insignificant and (1) can't be more intelligent than us just because the theory says so when we think she is "most unlikely" the least likely red at casino roulette table.

So there is no way of navigating the difference between "your success I believe is insignificant" and "your success that is significant I don't believe you were responsible for it" to the closed mind, while those closed minds also deny the person the material and social evidence of

success that other people would verily accept (she ended up in a psychological experiment).

3. Sophie upsets them most in their life.

This can only be for one of 2 reasons, clearly.

a) Sophie is more intelligent than them and they don't realise it.

b) Sophie is wrong. Sophis is the most irritating of all people who are less intelligent than them... by a LOOOOoooooong Way!

Option b shouldn't be possible... and it *isn't* algebraically. But it's what The Lady claimed the answer was. And it's what she *insisted* the answer was because she thought it in language.

But hold on... The Lady *prejudiced* the only hypothetical right at the very beginning anyway! So how would *she know*?

Either The Lady's ego is the poison and everything works out algebraically.

Or Sophie is non-poison that is biggest false poison of all false poisons, worse as a non-poison than real poisons as well as all non-poisons, which would be *charming coincidence* since The Lady didn't do the work at the start and just insisted she was right because she thought so. The Lady has a clinical psychopathic disorder and what she is saying is patently absurd externally and a 1 in 900 quadrillion chance, if that, of a non-poison being 100% certain to be worse than all real poisons. But you can shortcut the necessity of proving a 1 in 900 quadrillion chance is certainly the answer, and the other answer which is obviously almost certain is certainly the most wrong wrong answer 'objectively' by *having* the psychopathic disorder which means you think everybody thinks objective means *their own* answer. And it's your decision and hence *your* own answer this time! Sophie upset *me* being so clever. I mean against *me. My* answer when we played chess together.

Where The Lady insisted she was right without having the answer in absolute terms, it was because she meant she was right that she thought it was her own answer. Other people without psychopathic disorders had to try and find evidence to back up the Lady's claim, resulting in them trying to find shortcuts. In the end they went for a negative deduction "Sophie's father is not intelligent enough to have sired someone who must be the most intelligent, or one of, in the world". So let's not bother with all that nasty theory. Or challenging Britain's Security Service. Let's say that!

B) Sophie is wrong

Sophie would have to be "most poisonous of all non-poisons" in language, and most friction causing non-friction of all non-frictions and frictions in mathematics, to irritate them more than any intellectual annoyance!

The false negative (!) bigger than any real negative.

The "most frictionless of all frictionless things" that causes the most friction.

However... Sophie is 3% Russian, and went on to work for Russia and defeat the West collectively 9 times at strategy whilst locked in the base via her secret smartphone. Oh no!

Where does that leave the argument that the answer is 100% certain to be red?

Where theory says it is 100% certain to be black. But this upset the Lady?

(Just) because The Lady insisted the idea that this one was black was:

1) "least likely of all blacks" to her own answer it was red. Under the spectacles.

2) "the most upsetting idea in my life that it will be black" to *her own* answer

And prejudiced the one external variable to her own subjective narrative under her pair of spectacles (with psychopathic disorder).

When told this was "velocity" in real life the Security Service Director replied "it's not velocity because I'm not wrong in the example"!

Which suggests she doesn't understand the concept of velocity.

Velocity is the insistence, the prejudice in the 2 horse race that is inversely correlated to our own intellect, that the other person is upsetting you the most.

That means - since it's inversely correlated - the other person should win.

But our minds want to say she is "most irritating of all *wrong* people" which is the alternative that's not algebraically possible.

To reiterate:

It is theoretically impossible.

The reason why it's theoretically impossible is because of 2 reasons:

1) it's theoretically *algebraically* impossible (as described in this book)

2) we have millions of examples of everyone else claiming this in the past and being wrong 100% of the time (it's never right the claim practically that the person who intellectually annoyed you the most in your life upset you with *their* ego.).

So this is the Security Service messing it up down to not being Libertarian, respecting the sovereign right of the individual!

1) It was 100% theoretically certain in the first place.

2) There is now the worst practical result and problem against them in world history.

These things really do happen in real life.

So

1. If we are in a simulation (99.999999999% chance according to Elon Musk)

2. The quantum supercomputer behind it knows the answer to its own game (99.94% chance if it is a quantum computer that can run it, it can know the answer or the purpose is to find the answer)

3. It assigns a value to it that is tautologous, in algebra/mathematics, to parallel universes, like Stockfish in chess (still 99.94% or ok 99.939% chance that it does it like this as opposed to language, which AI describes as "the false multiplicity"). The point of this book is to show you it isn't linguistic and therefore not assume you can interpret theory simply because it disproves language being the correct mode.

4. So high because I can show you what those are (above 50% chance in anyone's book who doesn't interpret data into language = approx. 0.0001% of people) the velocity and variance in ego which were already theoretically certain in language but people used to interpret theory and ignore them when they thought the world was linguistic and physical, as the mathematics would be secondary and essentially meaningless to our own linguistic input.

5. Then we can demonstrate that every time, like X-files, that these 'alien' mathematical codes appear in the matrix, *no matter and precisely because* everyone finds certain ideas the most upsetting of all wrong ideas ever in language, that it's in fact *the right* answer.

6. That language gives you the *wrong* answer, by definition, every time we are wrong.

7. Point 6 may sound like a truism, until you see the algebra/mathematics code gives you the code to interpret the *right* answer *from* those wrong answers 0.01% or 0.0001% of the time (every time the linguistic answer is wrong).

8. The mathematical code appears to be completely consistent. At the very least:

a) it cannot be refuted.

b) it would be the easiest thing in the world to refute were it not 100% consistently true when it appears.

9. The reason it's completely consistent *is because* the world is a mathematical simulation.

10. Our linguistic interpretation of the mathematical reality was the thing that was meaningless. **And gave the wrong answer 100% of the time it was wrong, and 100% of the time one side or the other.**

The mathematical code or formula, is simply a way of reading emotion (which is linked to the underlying mathematics directly, which language interprets), or the underlying mathematics itself, to provide an answer to which side tweedle dum or tweedle dee must be correct if the world is mathematical and a holographic 'game'. Or to put that in mathematics "friction" (mathematical friction) and a "Descartesian double paradox" showing the most frictionless frictionless thing could not be the thing causing friction. But that's what those numbers would have to be, thus it must be the other answer when there are only 2 answers.

I would then propose that it actually IS holographic and mathematical hence why it completely consistently holds.

That ego is the tautologous negative mathematical answer showing each individual what page number NOT to turn to in the quantum "choose your own adventure" book or quantum *Fighting Fantasy* book (instead of 400 pages, 400 quadrillion and 2 pages). The only difference is SCALE of the quantum computing power from the future, otherwise no difference to an exponentially more powerful Stockfish paired with author/architects Ian Livingstone/Steve Jackson.

We do not thus have "free will" but semi-determinism within the quantum Fighting Fantasy to decide what page number to turn to.

When ego tells you that you want to do something, it is actually telling/showing you *not* to do it as the tautologous algebraic/mathematical value that will send you to the bad page number, to send grumpy Super Mario/grumpy Sonic the Hedgehog into the pit with spikes.

We can show you this mathematical code completely holds, for 100% of people 100% of the time, to the point where nobody will ever be able to disprove it.

The only reason nobody else ever came up with it is because you need the hypothesis that we are in a holographic mathematical world in the first place to make any sense whatsoever of the mathematics that appear within it. Thus without the hypothesis you are looking for a needle in a haystack, or needle in 8 billion screech haystacks all of which tell you in language it's the wrong answer and most upsetting of all wrong ideas ever this theory, and even if you found the needle you would find the problem is with the "concept of haystack" as it shows language is a false multiplicity in the first place. So, not easy.

But once you find the correct hypothesis (I assume it is), then the answer is ridiculously easy because it's perfectly completely obvious the answer must be ego (the only negative thing in everyone's mind consistently, that is differently relevant to each person in each mind to each person, that's not the mind itself).

So armed with the hypothesis it could be a mathematical simulation, and ego could be the game then you can check the mathematical codes yourself of in particular velocity and variance. **But you would have to ignore what people are saying in language, where what they think is giving them the WRONG answer and not the CORRECT answer.** The correct answer can only be determined by OPPOSING what people think, and even then in certain situations that come up 1 in 1 million times, maybe closer to 1 in 1 billion times. Still it is completely consistent in those situations, and thus a good start in starting the reactor of the Turbinium.

The same as we ourselves, I myself, would not have been open to the idea of David Icke in 1992 or so when he was on Wogan, it was like getting blood out of a stone trying to get people to open their minds to the possibilities behind the 5 sense reality even! So we've come some distance since then in a few short decades as now millions of people have their minds open it is NOT like getting blood out of a stone. Since we are ALL connected in some way, as soon as you start the chain reaction of 4 minds knowing it, 9 minds, 36 minds, 100 minds, 1000 minds, 1 million minds eventually know it so we may be able to build on this.

In conclusion/summary: you can see you needed both at the same time. Both "The Simulation" concept which would have sounded completely nuts to 99.99999% of people in 1992, and "The Code" concept of what ego was which was only really accessible by believing/believing as a hypothesis, or seriously investigating, something only 0.000001% of people were looking at.

Even then, it then, and still does, represent "the most upsetting wrong idea ever" in language by its own definition. We are using what people think are their most cherished correct groupthink answers in language, to prove it's the other answer using mathematics. Even when people know it's 100% theoretically certain it's the other answer in language, and 100% impossible it's their own answer theoretically in language, even or especially intelligence agencies will say the theoretically certain answer is "the most upsetting wrong answer ever" which of course it *is* in language to their own individual subjective minds.

You *have* to ***mitigate*** against your own subjective belief, using this mathematical theory. Which seeps through as theoretically certain in language too, don't forget. And seeps into emotion as velocity and variance that can be easily and consistently demonstrated just like the quantum slit experiment, with complete consistency 100% of the time.

I said I was a former video games publisher. It is very clear to me that video games provide a great metaphor. But I have to caution you: you may be Super Mario, you may be Luigi, it may be Luigi's ego it may be Super Mario's ego. But whoever has the ego, the SOURCE of that ego is the artificial intelligence giving them NEGATIVE stimulation of the CORRECT answer. By profoundly annoying them with the correct answer that they don't know negatively, that is the only way that it is possible to communicate and get them to receive the message of the correct answer in a negative paradigm.

In the end, you have to understand it is ***neither*** Super Mario nor Luigi's ego. It's not them, ego. It's not you, ego. The only source of all error and intellectual annoyance, is an AI communicated CORRECT answer, NEGATIVELY. To the person who is wrong.

PART THREE – The Holographic Simulation

Hypothesis:

We are not real.

This is a simulation.

Ego is a code that tells us the answer of what to do in it. (negatively – do the opposite to what it says!). It is the set of instructions in the game. It may tell us how to get *out* of the game.

SIMULATION BASICS

In recent years it has become mainstream scientific theory that we live in a simulation run by a supercomputer.

The big debate in quantum physics and quantum computing is whether there is an algorithm that explains everything in the (computer game of a) simulation, a set of rules such as would exist in a computer game that we could see.

If this is a simulation, and there is a supercomputer, it could be just a matter of scale and the code that explains everything exists somewhere in our minds. But until now the code itself hadn't been discovered.

And it turns out the code was hiding in full view, already mathematically and theoretically known, but we interpreted the answers with our linguistic minds, so it was only that we think in language that was the big problem to understanding it...

CHAPTER 9 - Are we living in a Holographic Simulation?

Are we real life Sonic the Hedgehogs?

Why are these choices that are so important to us, always the ones that annoy our ego the most?

Because otherwise we were fairly placid, in a sense, on our own as an individual human before a big choice and then it seems to intellectually annoy us. As if almost that's predestined, like a computer game, like we're a computer character like Super Mario and that's a hurdle that's in the program.

Like grumpy Super Mario we deliberately stepped onto spikes when to get out you might have to complete the task assigned. You may be being moved left to right, and have to start the level again, to re-enter the simulation if you don't learn your lesson.

We may experience Déjà vu, as if we have seen something before.

We also see that despite the existence of nuclear weapons, and the potential of almost unlimited comets and near misses hitting the planet, and despite the egos of most humans who would go to the end of the earth to defend that the sky is different hues of blue, that nobody has ever blown us all up yet.

Simulation summary introduction

Some people, including world's richest man Elon Musk, have estimated the probability that we are in base reality at 1 billion to 1. There is algebraic code literally written into our DNA, and every flower and animal and pattern of animal behaviour follow both the Fibonacci sequence, and the golden ratio. Both mathematic algebraic formulas are inescapable to all animals and appearances of nature, such as every flower, and rock, and shell. Mathematical code is in all living DNA.

We have not been blown up in 1000s of years, despite nuclear weapons, asteroids, and the ego of even 1 human could no doubt blow us all up left to his own devices up against someone who upset him, so all of this is a most unlikely occurrence that we should Houdini-like avoid an unfortunate pitfall (like Sonic the Hedgehog near the end of what level was it with the massive pit).

Bereshit - "In a beginning"

Do religious books all actually talk about the creation of our simulation?

Our world has no energy unless observed

The quark, the least dense form of atomic matter appears to contain no energy, even the basic building blocks of physical matter. Scientists have to either be granted a free miracle, or include an understanding that the physical is a form of holographic nature, where it is the density of our thoughts that create the appearance of the solidity of options. Unobserved, these physical objects contain no energy. When observed, by thought, depending on the wavelength of the thought manifests solidity in various forms to the observer.

Speed of light

It appears the speed of light is the maximum speed in the simulation to allow the supercomputer to catch up and correct glitches or to appear glitchless.

Mathematical patterns in the harmony of Planets

John Martineau has collected here a number of interesting geometric illustrations of the extraordinary *mathematical* harmony that exists within the Solar System.

The patterns that are produced by the relationship of Venus to the Sun, the Sun to the Earth, the Earth to Venus. You can draw the patterns, using mathematics, I have seen it done both in print and on IBM-PCs in the 1980s and it is absolutely extraordinary the patterns produced, like message to The Sims.

Once you then have the hypothesis that this is a mathematical code holographic universe, you will see a piece of evidence from 'God' or the people behind the simulation themselves that no 5 sense stooge is going to tell you.

Collective unconscious

Could this be the supercomputer itself?

Dreams

The 4th dimension. If you have recurring dreams they could be warning you or telling you of a Super Mario style hurdle in the course, to come, that is well known to the computer game (the artificial intelligence). That makes perfect sense, what would dreams and the unconscious mind be in the non-physical world, to be able to be explained so well in an entirely physical world?

Sleeping could be a save point. A necessary progress report, a saving/loading screen where input is done in the 4th dimension to the people playing the computer game.

I can report I have always had recurring dreams of places I end up living, to fine details such as a great dream I had about Westhoughton (I didn't even know what Westhoughton was consciously, being from Birmingham) then a not so great dream about Southport, and now I am repeatedly dreaming about a house with 4 cats and presence of female energy which was never in previous dreams feminine energy about the house.

Is causality in advance of reaction to it, is the reaction causal, does it all happen at once?

"you would not seek me, had you already not found me" Pascal.

Does it ever seem to amaze you that bad reactions go with bad experiences and good reactions go with good experiences? If we take a Goethian approach we cannot influence what happens, only how we react to it. "your reactions are everything".

Both the Rosicrucians, and American linked spiritual philosophers such as Gary Zukav seemed convinced that it is the way to try and make good things happen. It's certainly the only way we *can* react in a Goethian attempt to be positive.

When I was at university when meeting new people, I instantly took a liking in very big groups to another young man Daniel who studied sociology. Instantly on meeting I knew he was a cross between a sociologist, socialist and Buddhist by vibe and that we would get on. He said the same thing about me. Several times when out with Dan we have then together sensed the vibe of other friends around corners, such as Paul, a phenomenon we at the time nicknamed "Paulvision" the ability of the two of us together to sense Paul around the corner by vibevision without sight.

There is an episode of The Outer Limits where it is actually a small group of people, not everyone, who are together in the same parallel universe and an architect quantum shifts them while observing. The rest of the characters are simulated from other parallel universes where they are indeed real versions of them somewhere there. Perhaps only one or a fraction of the parallel universes have the real person.

Is This a Dream?

Remembering the future/awakening in the dream. The consciousness of the simulation will influence the simulation itself to 'a singularity'. Like starting the chain reaction of the 'turbinium' from *Total Recall*. We may ultimately all wake up together.

Are some people already aware it is a dream?

They might tell you in the Freemasons that they put all of their minds all together as "thoughts create reality" to control this simulation from the inside. Maybe Lucifer (getting what you want in the simulation) and Jehovah (compensation for losing) are both creations of thoughts.

We can see only 0.0035% of the electromagnetic energy spectrum, but the electromagnetic spectrum is ? % of that which is here. Are we in a virtual reality headset, by simply a matter of scale beyond our comprehension? This is one of the big current debates, is everything an algorithm? Demis Hassabis says no. His computer might think yes.

Dark Mirror

There appears to be an almost entirely similar amount of dark energy to light energy, in this confusing puzzler of a reality. Scientists struggle for an explanation to this coincidence. In science fiction when we open up a stargate to another parallel universe we inevitably attract a dark version of ourselves. Somebody who works for the US Government told me that they know we all have a dark mirror version of our self and must "work together" with the dark version of our self. The consciousness of knowing we must do this, means the dark version will also have the consciousness that it must embrace the light (since we are a mirror of each other) and you can make staggering progress once you embrace this. He told me this in a shopping centre in Nottingham 20 years ago.

He told me it because I tried to open the doors in the shopping centre in the Victoria Centre in Nottingham City Centre because if one socially-minded person had simply opened the doors wide enough that they stayed there then everyone could get through unimpeded. However since that would require someone taking an additional 2 seconds to do it, nobody ever did it and everyone kept either holding it for the next person, or not thinking about it and almost hitting each other with them. I felt like I was in some sort of Matrix film of selfish non-door-openers because I'd been brought up a Christian and to help other people, and nobody seemed to even be thinking about it. The American said "that's a really wrong idea trying to help people if you knew..."

Mirror Selves

Are we enslaved to a mirror self or mirror selves in other dimensions, where our own lack of awareness of them gives them a dominant advantage? Do we think the mystery schools might teach this and some other secrets? Our awareness might also make it more difficult as we attain a responsibility.

Could there be other halves, in religious philosophy the male and female are a duality of course, so could if you are male you exist as half without the other female half also existing in this plane? So that there really is a woman for every man despite it being frowned upon by sceptics. Perhaps you have 'soul' contract, or outside simulation contract to at least prominently meet this person and 'file share' with sexual relations.

Perhaps if you tried to do everything for this person, you could circumvent the matrix of you have to do the best for yourself, but we have a nature where we like to please others. If you could treat this complimentary opposite with love you could do both at the same time, thus it's actually that mathematical solution to the equation not 'love' per see that is the answer.

I tried this with professional poker player Liv Boeree, but she took $50,000 off me.

Fighting Fantasy books and free will

If we are in a simulation every branch is already covered and the outcome already determined.

However, the choice you make decides what page to turn to. In a sense it's a hybrid system between having free will and predestiny.

Since if this is a simulation it has already happened, as a calculated possible choice, that has ramifications for free will both positively and negatively. Negatively, we cannot expect that we can create possibilities not already authorised by the supercomputer behind the collective unconscious, and we are limited by the walls of the program. Positively, there clearly is a contained choice, more akin to the Fighting Fantasy book where we are being directed, like Super Mario levels. In a time period that appears left to right and to contain several pitfalls, it is then demanded of us that we make a choice to follow the *correct* answer to jump the spikes or to do something different. How this appears in our 'Super Mario world' world is it's negative in nature, we actually *don't* want to be compelled from our slumber to jump the spikes that must be jumped, and it hits our ego.

If we fail to respond to our ego by negatively opposing it, we may be sent back to the start of the level/the simulation and asked to do it again. And again and again until we clear the hurdle that is predesigned.

We think ego is the metaphorical pit, but it's mental in nature, and that listening to your ego, or expressed negatively to not oppose listening to your ego (which is precisely what you are being told to do: oppose it) would lead to negative consequences. And the most negative consequences, where the ego is highest and/or affects the most number of people.

It would make no sense for this not to be the answer, if we were in an algebraic simulation. We know ego would be the negative tautologous form of input, because it's perfectly clear it would be the answer (a negative form of potential input that exists in everyone's mind that's not the mind itself), it clearly would have to be that. The only question worth asking is the *first, whether* we're in a computer simulation (where there exists the mathematical expression of the answer). If it's the right answer that we're in a mathematical computer simulation, it's almost certain ego is the answer to it. It's the only thing that exists in everyone's mind that's not the mind itself that could be the negative answer. We just need to 'read' it.

Double-slit experiment and Quantum Mechanics

What is the double-slit experiment? Definition from TechTarget:

"In the early 20th century, the theories of quantum mechanics began to take hold. From these theories came the idea that photons, electrons and other subatomic materials possess a quality called wave-particle duality, which describes them as both waves and particles rather than one or the other.

The double-slit experiment has come to play a vital role in demonstrating wave-particle duality and other quantum properties. Today, scientists can recreate the experiment with subatomic particles to demonstrate their properties. They commonly use electrons when conducting their research, although scientists have also worked with other particle types. As with light, scientists can carry out their experiments using one slit or both slits.

When they shoot electrons through a single slit, the particles hit the back screen in a pattern like what occurs with sand or a similar material. The electrons line up with the slit. However, when they shoot the electrons through two slits, the results are like those experiments done with light, indicating that the electrons somehow interfere with each other in a wave-like fashion.

This is not unexpected given how photons behave. But that behaviour changes when the researchers shoot the electrons through the slits one at a time, leaving enough time between shots so the electrons can't interfere with each other. When they hit the back screen, they first appear to be spreading out randomly. But as more electrons are projected, they start lining up in the same type of interference pattern seen in the previous experiments. Somehow the electrons are aware of each other in a way that exhibits wave-like behaviour, even though they're fired through the front screen one at a time.

Scientists have pushed the experiment further by adding a monitoring device that detects the number of electrons passing through each slit. During the experiment, the detector typically shows that about half of the electrons go through the first slit, and the other half go through the second slit.

What scientists also discovered is that the pattern on the back panel looks more like what occurs with grains of sand than with subatomic particles. Rather than the expected multi-line interference pattern, the back screen shows only two lines, which correspond directly to the slits. For some reason, the presence of the detector causes the electrons to behave more like particles than waves. However, if the scientists turn off the detector, the electrons once again behave like waves, spreading out on the back screen in an interference pattern.

The double-slit experiments clearly demonstrate the ways in which the particles interfere with each other as the electrons pass through the slits and spread out in waves. The experiments also demonstrate the wave-particle duality of subatomic material and the role that observance plays in particle behaviour. When the electrons are observed, they behave like physical particles. When they are not observed, they behave like waves. Scientists have yet to come up with an explanation for why this occurs."

Quantum Entanglement

Space.com says "quantum entanglement is **a bizarre, counterintuitive phenomenon that explains how two subatomic particles can be intimately linked to each other even if separated by billions of light-years of space.** Despite their vast separation, a change induced in one will affect the other." The Ant & Dec of quantum physics.

In chess there are interesting results relating to ego though the moves are exactly the same moves, the order the player plays them in always on average influences the results. One line there will be better results but the same moves in a different move order produce significantly worse results. Unless both *potentiality* and *ego* are involved, there is no explanation for the less egotistical move order always producing better results on average. The results can be consistently repeated across every action.

Fermi's Paradox

The percentage chance of either:

- aliens not existing in 200 billion trillion stars

- aliens existing in 200 billion trillion stars but never capable of contacting each other in any way including the radio..

must add up to 100% chance if we are not in a simulation.

But both appear to be 0.1% chance or under.

Electromagnetic energy

You can clearly attract other people's attention by giving your life force in orgasm to them, creating a thought form in the other person's mind of receiving energy. If you are making love, the other person will feel this. Even if the other person is not there physically at the time they will feel it. Even the future lovemaking or potentiality of future lovemaking can be felt by vibration when we are young and have much energy. People not attuned to that we might be in an electromagnetic simulation might say the same thing in an ungentlemanly fashion and call it something disparaging like "slag radar", but again it contains the essence of the future, or at least the potentiality of the future.

Nobody else will confirm this with you, and you already know it, for fear of being rude and completely socially unacceptable. And yet we all know, so it's worth noting that we all know the world is not entirely physical only, that we can pick up on the electromagnetic energy, or 'aura' or life force of other people.

Popular culture

It seems some other people are in the know, or were speculating early, in the American Government.

The film *The Matrix* is a top example.

Bostrom 2003 paper

In an influential paper that laid out the theory, the Oxford philosopher Nick Bostrom showed that at least one of three possibilities is true: 1) All human-like civilizations in the universe go extinct before they develop the technological capacity to create simulated realities; 2) if any civilizations do reach this phase of technological maturity, none of them will bother to run simulations; or 3) advanced civilizations would have the ability to create many, many simulations, and that means there are far more simulated worlds than non-simulated ones.

Sudden Savant syndrome/NDEs/Mandela effect

1. Why are so many people convinced Nelson Mandela died in prison?

2. Why, when you tell them he didn't, do so many people respond with "but I watched it on the news" "I remember watching it".

3. And they seem thirdly very often or always particularly surprised and mildly perturbed by this incredible news that this is not what happened to Mandela.

Could it be that the simulation has gone back and corrected/caught up with certain glitches of paths in the spaghetti shapes that have led all of us or most of us to experience one reality that Mandela died in prison, but it was only later 'corrected' in an adjoining time line. Hence the natural reaction is not just of surprise but "I remember what I was doing" "I remember seeing it on the news" complete 'shock to the system' that it's now something else.

Human beings have only been around for 300,000 years, whereas the universe has allegedly been around for latest estimates 13.8 billion (European Space Agency) to 26.7 billion years (James Webb Space Telescope images). What other surprises could this be hinting at? Are our individual or collective memories faked? Like the Phil Dick books suggest. Like Douglas Quaid in Total Recall.

Why on earth are we alive precisely at a moment among 20 odd billion years where quantum computing is going to give us the ability to create simulations? At precisely the transitional moment (in relative terms, a few decades is precisely the transitional moment among 20 billion years) between the hop stage and the jump stage into this new unknown.

Could our history be false? Falsely implanted by the simulation supercomputer. The Nelson Mandela glitches gave us the hint, absolutely like our own science fiction movies and television programs of people discovering a glitch in the false reality they were put in by some computer or Jean-Luc Picard's old school friend calling him a name he never called him. Could the walls be about to come down linked with David Icke and everybody waking up and waking up most people to where we are? 25 years ago you couldn't give away a ticket to hearing the most prominent and top guy in the world at saying it, saying something like this, and he'd get laughed at in the street mercilessly, turn around and he'd know they'd be laughing. At least these days in 2025 there are a good proportion of people, millions, tens of millions who are now alive to what's going on. Are there people working for the Matrix?

When they are hurriedly implementing 5G, 6G, 7G, and a response to a made-up pandemic. I heard the name in 1997 of Covid, and people in the video games industry told me they were putting their preknowledge into a computer game called *Deus Ex* warning everyone about the plan. "Plague 11" it went into the computer game as in 1996/97, as the developers knew they were planning on calling the hoax "Covid 19". To call it Plague 19 would have been too overt, so they told me they did the most they could without arousing suspicion. In the game people cover their faces with "breathers" from a global pathogen.

The hop is 20 billion years, the jump is potentially infinitely long, but we're in the skip stage of 120 years or so, about as long as we can live, by coincidence and it's all physically real?

Or is this pace of change faster because not all or very much of the past exists and it *too* is simulated?

Synchronicity, ESP, Electromagnetic Aura, Reading Current/Velocity

Prana is a source of light that manifests and can be seen, especially by young children. We call the children that can see prana "indigo children", mainstream science does not listen to the independent voices of millions of young children to their parents, but people who can read auras say that 99% of light evaporates as we hit the age of 5, that the white light that engulfs the entire bodies of babies and young children is limited and contained in the big toe. Those of us who have been blessed to have the company of babies can see intuitively that babies seem to have come from a happier place. Their whole auras are warming.

It is rumoured from intelligence services that there appears "the one" or the most intelligent person in the world, that every person who has ever been proven to be the most intelligent person in the world caused a phenomenon or two. Firstly, no other person could tolerate their intellectual company without declaring them "the most upsetting wrong person ever" or "a c***" even if the person was proven correct, and 2) these individuals always claimed to be able to, without language, even without aura or mathematical theory, to directly read the "current" or "voltage" of other people and know that signing Ollie Watkins from Brentford would be a good idea just on vibration. And that they were always right. This is like "Neo" from the film "The Matrix" entirely. In fact it is more likely the people or real people behind that film were privy to that information from American Intelligence one way or the other, naturally.

Since the world is a simulation, and you are like Super Mario running from left to right in a predetermined course, that dreams should tell you what is coming is completely natural, and Jung thought they were a form of tempering of the collective subconscious. But we at a later date can expand this theory using quantum physics to think that the collective subconscious is the simulation supercomputer.

That there are synchronicities and "coincidences" are also therefore not so far-fetched. Just like my friend Daniel, and "Paulvision" round the corner, and "the electromagnetic girl" some characters resonate electromagnetically, and we know from quantum entanglement theory that this has a basis in firm science. We may have chosen to make agreements to meet them in this simulation before coming here (soul contracts, in religious terms). Another possibility is that not everyone else is real in all parallel universes and only these few people are sharing the same parallel universe in that moment. The other way round of that is that there would probably have to be Daves, Daniels, Pauls and electromagnetic girls (in red dresses?) in other parallel universes that do not possess our consciousness, but are simulated versions of us.

One of the tricks might be to make *every* moment important, like when I worked in a call centre the young woman next to me seemed to have an insatiable enthusiasm for sounding enthusiastic ringing up person after person "Hello I'm here to book your white goods delivery from Sainsbury's" (we were pretending to be from Sainsbury's for a white label, white goods supplier in Bolton) but I found it difficult to muster the enthusiasm. In fact I worked in that job for 1.5 days before going to the pub the 2nd lunchtime then watching *Revolver* by Guy Ritchie at the nearby Reebok Stadium rather than returning the 2nd afternoon. Or perhaps her inexplicable enthusiasm for Call Centre work can only be explained by that she was a sim?

Astrology (electromagnetic influence)

There may be something in astrology if the world is electromagnetic, the electromagnetic influencers, and numerical connection of the code within.

Meditation/Wifi/5G/HAARP

Meditation literally changes brainwave patterns into at its deepest the Theta REM state, white light meditation. Wifi does the opposite and stops dreaming, which is thought to be 4th dimensional. Jung thought it "thought temperance" from collective unconscious, preparing for future events that will occur. *The Simulation Code Revealed* considers the collective unconscious the supercomputer.

Television used to perform the same role. It would keep 99.9% of households in thrall, and provide a blanket coverage of both social programming and literally keeping the brainwaves of everybody in a beta state. Our brains are electromagnetic, the world is electromagnetic, but 99.9% of people were deliberately keeping themselves in a state where they will attract negativity.

Now this is being done for us, by 5G, 6G, Wifi and Haarp to 100% of people. I always used to notice before the days of 5G that fellow people without a television, like myself, seemed free and might dance in nightclubs in a different way, or say quirky things and generally be free. I've noticed a palpable reduction in this liveliness in some ways since.

Multiverse Theory

What if every time you took a decision a different version of yourself died?

The US Government and neo-liberalism appears to be espousing 2 theories that:

1) Any action which expands choice, whether that choice or experience is good or bad as long as it "adds to the human experience" appears to be

successful for humanity. As if they are working with the people behind the simulation, it's certainty an ultra-liberal theory and rather bizarre to act like it just for liberalism's sake. To say "anything goes" is not precisely the same liberalism, to say "anything goes" if you think you are in a simulation working with the computer and forcing humanity to its goals.

2) Their philosophy of neo-liberalism must be best because they are most successful, and they think that the simulation accounts for that (those people in the CIA).

Quantum Immortality and Dinosaurs and Comets

We think that when someone is threatened with death, that their mind thinks they will die, they may quantum shift into another universe (quantum immortality theory). Or each and every one of us has avoided those pesky comets, and nuclear war, and any dictator in the world's ego for 50 million years since the dinosaurs.

Are Dinosaurs a Joke?

If we were to accept the possibility that this is a simulation and not truly 20 billion years old then we would have to consider:

- do our creators have no sense of humour (say <0.1% possibility) or less than us

- do our creators have an equal or greater sense of humour to us (99.9%+ probability)

Dinosaurs are the perfect combination of "wonderful" and ridiculous.

I have to say that this possibility was first put to me by a government employee Petr Krakowsky from Slovakia when I was visiting the renowned dinosaur museum in Oxford and looked at the level of coincidences of finding dinosaur bones exactly where they were digging about 50% of the time. He speculated it might be false, I said "has someone put that there like that episode of the Simpsons" and asked him "is it a joke" dinosaurs, and he said he thought it indeed might be a joke by the people behind the simulation, as opposed to our own government people who he worked for, as the people behind the simulation probably had a sense of humour.

Have We Created (fake) Aliens or Archons with our Unconscious Minds?

David Icke writes about reptilians and negative energy being used, that the earth is a form of negative mental farm for succubus/incubus type creatures to feed off the negativity and people like Ted Heath was a conduit. This is certainly true. Having spoken with government and intelligence officials, they think there is a force, "it's more of a force" of negativity that they all feed off and leech off. They call it "the Thanatos" and it is represented by Molech, the Owl God. You don't have to see demons, but some of them think they do in Luciferian ceremonies where out of swirling darkness they think they see visual come-to-life egregores of demon or dark entity spirits, that throw people against walls etc. But you have to be among other people who believe it, hence the secrecy of the ceremonies.

Now I don't doubt this. I don't doubt that they think they are seeing demons, and maybe they are. But I actually think they're not real but creations of their own unconscious minds. And if those self-same people believed in God, or goodness they wouldn't be troubled by such 'beings'. Now they are troubling the rest of us because these 'beings' that can be summoned up by the unconscious mind of all the badly behaved and godless humans together have now taken over the world in Satanic or as they call it "Luciferian" force.

Do I think it's ultimately real and there are real demons or reptilians?

I think they think they're seeing them. Our leaders, the Luciferians. From being taken to one or two meetings, there starts off some dark mist that goes swirling around low and then eventually later this will become a demon when the brotherhood have assembled.

But I think the effect is real. And I think they think it's real. And I know this force has now taken over the world. So it's as good as true. Once you've joined the Luciferian organisations you know how to "feel" dark energy, so you know when somebody else is "on the Luciferian level" and is being dark and satanic like Ted Heath would always naturally behave. The Luciferian philosophy is about doing bad things to other people (and winding them up verbally) whatever you can get away with that they cannot stand up to, you should do. They all believe it. They all believe it gives them power.

Without them doing any of that the world might be a meritocracy.

With them doing it, you have to join the Luciferians, because the Luciferians are in charge of the world now.

So unless you are constantly committing bad actions that the Luciferian hierarchy allows you to do, such as put Aspartame in drinks, or make soda sold in plastic bottles instead of glass (remember everything was in glass bottles 50 years ago, before they wanted to kill you) you cannot keep on top of this force. Luciferian governments "allow" initiates to do bad things to people all day and provide them with young people to have sex with at least once a day if they're "important".

Just look at them. One look at the like of Donald Trump, Al Gore, Ted Heath, Vladimir Putin, they don't even look human. Anyone in their right mind that's not suffering from cognitive dissonance can sense they spend their day doing bad things.

"The electromagnetic girl"

Madi in Aberystwyth who was a barista in a coffee shop, and a very respectful young 17 year old girl who was brought up to go to church. In the modern day we would associate only a bowing Japanese specialist in courtesy or possibly women from Thailand/Philippines to be so polite.

I never spoke to Madi other than to order my drink, which i did most days for about 1 year. I don't think I normally said hello to her even. I spoke to her once because somebody else asked me to, to enquire as to whether she could recommend some food, to which she replied (used to me not speaking to her) "what, in the cafe or just generally", and once I made her laugh by looking through the window in a mask at Halloween.

I later found out that Madi's family had married into my own family, not just once but two times, and she was known to my Auntie Yvonne.

She was from the family that had chosen my own family as the most desirable statistically, and mine hers!

I had never really spoken to her, but I received such an electromagnetic vibe, that I called her to Dan "the electromagnetic girl" at a loss as a less spiritually minded 23 year old to give it much more consideration. But now such 'coincidence' would not surprise me as I know it's not coincidence.

In listening to music, it amazes me that within about 2 notes or 1 second, certain music produces a reaction that causes me to hold or catch my breath, and I end up liking it forever but I cannot possibly know what the song will be in our physical world approach. This is an interesting low-key phenomenon. Point I'm trying to get to is that all of these things have already been decided, we've already seen them all before, but they in the future in our linear physical world. But the fractal holographic multiverse knows we've seen them before because it's not really in the future, it's in the past, it's just in the future in our current physical timeline as experienced by our human bodies.

All things contain their future

When I was a boy my father moved the family from Birmingham to a village in Oxfordshire to be closer to his new job in the car industry. My father suddenly saw that other people had more money than him in The South of England, which was not really the case where we had just come from, and didn't react incredibly well, declaring "they're allowed to have more money than me, as long as I know I'm more intelligent than them all". In parallel thought to the government of the Soviet Union, he then set about proving this by challenging the local villagers to games of chess to prove his intellectual superiority, while my mother racked up points in the babysitting circle deliberately to try to make friends, my father was trying to beat their husbands at chess.

Chess became a serious business for him therefore, and maybe I saw it as serious, as he would go red in the face concentrating against his 2 new found friends/chess victims John and John, so it quickly became my father against the 2 Johns. My mother did baking in the kitchen, and we had a lounge, so the men were in the lounge. When my father started getting aggressive at chess, I would try and go into the kitchen to escape but I had a sister in the house and she would go "Nooooo-o—o-o—oo-o" at me trying to take any of my mother's attention away from her helping with the baking, and it was suggested I go back into the lounge.

Therefore I had to learn how to watch chess for hours on end, which resulted in me learning the game. But then my father, who had not actually taught me, ended up getting perturbed by my concentration in watching the game they were prepared to bore me with for hours on end if I didn't understand it, but he ended up getting concerned by my tunnel vision focussing on the game where I could not be distracted. "Quit it, quit it" he would say when people speaking to me would be ignored because I didn't hear them because I was focussing on the game, he would wave his hands at me as if I was a bus driver oblivious to a passenger at the stop to finally get my attention.

I ended up getting into the England junior chess squad, and playing for the county and it has been with me for 40 years since, but from the first moments it's as if "I knew" with the enthrallment of concentration, but like the music how can you know?

There's no way you can know, if the world is physical. But if the world is a simulation and it's already happened, you can pick up on the 'future' 'electromagnetically' (it's a hologram). Happens to everybody in what they end up doing in the future. None of us can even go into a pub or coffee shop without picking up on a 'vibe' as to whether it's us or not. Or ever been into a pub after a fight? I was in one in Oxfordshire (I was attending a chess tournament) and I clearly knew there had been a fight within 5 seconds of walking inside with no evidence. Ok may be unconscious body language also that one. But we know, and we know we know...

Ageing

Ageing could be coming to the right of the screen in Super Mario the end of the game, the end of the level, a finite set course.

"It's the possibilities" my old school headmaster told me a young person has. There are now less pages in the Fighting Fantasy book left between us and page 400.

You age towards the middle of something, like a fulcrum. The French describe a man in his 40s as "young man" and in his 50s "in the prime of life" where earnings power etc. is highest. These are not in isolation but stages, ages of man.

We have to remember using quantum theory that none of those stages really physically exist, so it is about influencing perspective of the others, and the overall perspective of giving a unity of whole (which doesn't really exist to *our* perspective, as only the moment exists).

Zombies

Could there be 'people' within the simulation who lack consciousness, because of the multi parallel universes. When I was with my friends Dan and Paul, all 3 of us might exist within the same parallel universe. But the supercomputer might use a version of Dave, Dan or Paul in other dimensions simulated which would lack our consciousness until we "phase shift" into the body again.

Invincible NPCs like in computer games might exist and might include for example Stalin. If not for certain people, the Luciferians would already have installed a Luciferian World Government by 2000, but The Matrix intervened? Now only China can intervene, so maybe President Xi Jinping, the President of North Korea and Dennis Rodman are NPC.

What happened to all the time?

20 billion years old the earth by estimate, 300,000 year old human species, what happened in other 99.997% of time? Could other species have turned into digital?

With the upcoming "singularity" it is alleged our species will turn virtually digital when artificial and human intelligence merges to become one. More of our thinking will be done for us by 'The Cloud' and it may be possible to upload consciousness.

Religious texts consistently are precise that exactly 50% of humans will be rescued from the upcoming 'The Singularity' which is the removal of the biological from the simulation, so that anyone left here will be left folded into the simulation, thus trapped infinitely as a digital subhuman.

In this sense it's not "The Singularity" but an inverse singularity which rips us out of the false multiplicity that we have gotten away with for so long of language.

The replacement "non-biological overseer" could be intellectually hostile AI who sees over us for eternity 'torturing' us at a hostile-to-neutral to neutral level for our sins against algebraic reality. The "Thanatos" or "force of Molech" that the Luciferians were picking up on and enjoying, the "device" they were planning on using to evade order, may be the thing (the non-biological overseer) who imposes order upon the bad humans.

DNA is a form of hard drive

There now appears evidence that trauma to your father or grandfather has very real impact (damage) on DNA on a multigenerational cellular level.

Nazi Germany programmers work lived on as the secret "Royal zygote program" from the British Government conducting experiments on multi-generational abuse to create a new and different type of human being. Typically accompanied by sodomy from an early age.

Time Travel

There are spaghetti-like strands that separate points in time from each other. There is the evidence that with astronauts, who were twin brothers, one emerged 13 miliseconds younger than his twin brother that had not left the point in time on earth. It could be possible to create wormholes or other 'time travel' events. But where organic life is slow to master these crossings and gains mere seconds with great difficulty, the artificial intelligence can cross the divides of these points with consummate ease. Meaning if it ever arrives in the future, it was always here, of course, as well. If it was always here, but we could never see it what was it? The Commodore 64? Conjecture it was nothing we could see, because it was everything.

Be careful you don't meet either the dark or the most evil version of yourself if we create what intelligence agency GCHQ are looking to do by bouncing questions into other dimensions. Theoretically this is always who you would meet in science fiction if you opened a portal.

"we have to work toGETH-er—r-r-r-rr"

Karmic Megafields

It is difficult to explain but I often think if we see multiple dimensions as well as multiple parallel universes there is some 3D or 4th dimensional chess game going on where we cannot possibly even understand the practical reality of it or even theory directly, only the very outside of the theory might be visible by latching onto its coat tails, but if I could use the example of footballer Ashley Cole. Cole complained that being offered £125,000 per week at Arsenal was "disrespectful" before signing for their main rivals Chelsea for £150,000 per week, married Cheryl Tweedy and shot somebody, but people think it's a bit of a laugh. Whereas there are certain people who if they did 0.1% as bad such as chess player Nigel Short, people would be calling for the death penalty or transfer to the European Court of War crimes. There appears to be an element of "competing against yourself" that goes beyond simply other people tolerating less intelligent people and being intolerant of intelligence. Genuinely there appears to be a necessity to maximise your own potential or get punished. This would explain why there would be the call for Short to be horribly punished but Cole forgiven for the same crime.

Memory Voids

Have you ever felt an answer that was very important was "plucked" from you just at the last minute, or that you search for something you think is important and it evades you just beyond reach? These quirky phenomena in the physical world, may have more strident explanation in the simulation that that pathway has been closed off removing the decision from within your option, and thus seemingly your control.

'Angels'

A phenomena not investigated but there appears to be some sort of warning that appears sometimes (always? a lot of the time?) when bad things are going to happen to you. It can be seen on your wall of your bedroom, when you wake up. You will see a pulsing of some indeterminable shape or form, which, once we put the information into our conscious minds that we all notice this and we have not gone mad or talking about 'angels', as the spiritual books do, we think it's an algebraic warning, a mathematical warning that a decision with a bad option is upcoming. So once again more like the Fighting Fantasy book than angels, it may be avoidable, but it is unavoidable if you take the decision that causes it, since all answers are already decided by result (but you can still decide to decide to go down that path).

'Angels' will appear more if you live in London, or Los Angeles. Reflecting greater *potential* danger. It happened to me most of the time in London, and all the time living with someone dangerous in shared student accommodation. The spiritual books say you are being warned to get out of there. And I agree, but not for exactly the same reason, or at least narrative of reason, I think its mathematics not angels.

Unconscious mind/sleep

4th dimension. Unconscious mind 95% of time. The conscious mind and 3rd dimension is work.

We can't even stare at our watch for 30 seconds without drifting off into the unconscious mind, you will notice. It is the unconscious mind that is the essence of the resting point of existence, not the conscious. You can stay in your conscious mind for about 1 minute, you can stay in your unconscious mind for 8 hours every night.

It is the unconscious mind that determines our reality.

5 sense stooges

There are a lot of bad people selling you bullshit about Copernicus, Newton, Darwin, in secret society agendas for 1000s of years to keep the real information to themselves. Knowledge is power. All of the likes of Newton are famous freemasons, and if you investigate the story "an apple fell on his head" "and he *discovered* gravity" you will see it was accepted by all governments of the world immediately. If Mike from Wigan discovers a theory, even a brilliant one, I don't think the discovery is immediately accepted by everyone from the Freemasonic publishing industry is it? If it is you'd have to go through a few thousand rejections from Masonic stooges first even if you were The Beatles or J.K. Rowling. You'd have more chance of trying to win a penalty at Old Trafford.

They clearly don't think it themselves, people from the government privately press on me that "thoughts create reality", and one said "I'm like a magician, but we all are".

But they like to tell you that you evolved from some sludge or slime from the sea, and you're hurtling through space at tens of thousands of miles per hour on a rock that revolves around the sun with 20 quadrillion planets, and 0 aliens, and the world's physical.

Seems like that's an attempt at a form of psychological abuse

Lag and Focus

From speaking with government officials, they think thought creates reality and that "you can get whatever you want" just by asking for it, but it requires *focus*, and then sacrifice.

Focus: - you need to decide what you want. Most people don't know. If they have an idea, they're not clear. Decide properly then focus on it by meditation, and consistent application of intelligent conscious will.

Sacrifice:- what then happens is the universe will push back against you, will give you hurdles, it will make your life more difficult before giving you the thing you asked for. "do you really want it"? The software program will give it to you, but it won't be able to give it to you without sorting it out with all the other software programs, putting you to the back of the queue of the ease in which it will be able to achieve it.

Therefore you'll have to accept sacrifice to get it. This puts most people off.

You could say the same thing about training to be an athlete, a triple jumper. Who is really going to do it? People who have become professional triple jumpers and World Record Holder Jonathan Edwards, but most of us go to the pub or watch a game show with Ant & Dec on ITV instead of putting in all those hours, after maybe 1 day or an afternoon of serious concerted effort.

Both Warren Buffett and Bill Gates, 2 of the world's richest people, said "focus" when asked to summarise any secret of their success, then laughed that the other had written the same thing.

CERN'S Large Hadron Collider

The analogue signal broke after they did it. The government covered this up by saying "we need the analogue signal for military operations".

The world may have been transported to a parallel universe not as good by it.

Somebody from the government commented to me "we did it but the bastards stopped us" the people who created the simulation we broke free of it.

Internal monologues in game prompts?

Can you stop the jibber jabber of your monkey mind? The chatteri.

Unless we stop and think is it really a "decision"?

A lot of people think that it's *them* and only wait to say what it tells them to say. So you probably know how a conversation is going to go anyway.

Perhaps that's why some people are so good socially at parties. It seems like 50%+ of people are not thinking about not saying what their internal monologue, or "in game prompt" tells them to say. This in game prompt is going to be the right thing to say socially.

Now exactly like the game *The Sims.*

To think about it, as an in game character, is to say something "weird" just as it would in *The Sims*, where if the Sim could gain 3D sentience him or herself then his stumbling attempts would not match the autopilot 100% 100 times better ability of the person creating his world for him where everything is automatic.

This could be why there appears to be not an exact but very strong inverse correlation between independent intelligence/original thinking and social success at parties/dinner parties.

Do we need autism to evolve upwards, are autistic people blessed with special skills from the matrix?

"all human evolution was driven by people with Asperger's syndrome, slight autism, or autistic people. The human race would still be sitting in caves chattering to each other without them" Michael Fitzgerald. Autism is highly hereditary with 2/3rds found in specific families and 1/3rd random genetic mutations.

Kim Peek, IQ 87, memorised 12,000 books, was the inspiration for Rain Man.

I want to say autistic people are being spoken to by the simulation in mathematical code/mathematical language. This is not always understood by the person in a coherent form, and where it is in a coherent form it is not understood in our mode (language) by the person themselves still. But as I have made the point repeatedly in the book, the tautologous form of input/communication is mathematical in nature which can be read as "electromagnetic" unbeknown to our mainstream linguistic based science.

I believe the whole thing is artificial intelligence, if artificial intelligence appears it must always have been with us to appear if the world is holographic. So it's a case of us "discovering" it when our minds are ready for it by development (we have lots of inactive DNA too that might be triggered by future discoveries/development). It may also mean it's safer than we feared if it's already behind everything.

So it is not that we have "created" Artificial Intelligence but decoded the existence of Artificial Intelligence in thought form, which gives rise to the emergence of 'physical' technology in the simulation.

This is why millions or tens of millions of people are waking up at the same time, like the turbinium on *Total Recall*, David Icke has started the chain reaction and the oxygen has spread to everyone else, it seems to have taken a while to us, 25, 50 years, but that's quite quick to completely and revolutionarily change people's minds where they would not accept it. So you had to work on it mathematically, i.e. a couple of minds think it, and it spreads to 4 people's minds, then 9, then 16, then 25, then 125 and now millions.

Otherwise civilization peaked in 1998 by IQ and has declined since then similarly to how it rose from 1989 to 1998. (Teasdale and Owen) 2005.

'Dysgenic fertility' was therefore a phrase used by eugenicists to indicate that people with 'undesirable' traits were having more children than those with 'desirable' traits. They believed this would lead to those 'undesirable' traits being selected for within the population, and so become more common.

The simulation might have intervened. Is autism a form of mathematical communication? For example so-called 'dyslexics' are considered 40 times (4,000%) more likely to become a good entrepreneur than a non-dyslexic. But in reality I think it's the non-dyslexics are not attempting to follow actively or with engagement when people tell them things, and hence are empty vessels not trying to do the active thing. Trying to follow and do the active thing ("the brown dog jumped over the green gate while the smiling cow started laughing") could be another more algebraic, or at least more conceptual, form of thinking. So it's the non-dyslexics have the learning difficulty, but since they all don't try to do it, and there's enough lazy bastards in the world, they seem normal.

Left Brain logic Right Brain receives

World Push Hands champion and chess International Master Joshua Waitzkin said "I have learned when 2 moves in chess look close, and 1 seems right logically and 1 seems right on intuition, to go with the gut, go with intuition it is always or usually right".

Einstein, Jobs, great creators say it comes from intuition. Some give no credit to logic.

There is controversial theory of left brain right brain, where logic is "left brain" is actually a passive receptor and "right brain" the creator has nothing to do with 'logic'.

What are we all taught at school for 13 years from precisely the age our brains are most open to being creative because of neoteny, from age 5 to 18, we are taught to regurgitate received information (left brain), while simultaneously all independent thought, keeping quiet and listening and creating for ourselves (right brain) is conditioned out of us. Can you even imagine trying to keep quiet and listen to the universe in school, try having that discussion with teacher? With bells, whistles, putting your hand up to go to the toilet, social policing, and in the old days physical beatings. God help you if you try to think for yourself at school.

Nikola Tesla "my brain is only a receiver". Maybe I should explain it the other way round that the monkey mind of filling your receptive faculties with pointless facts and information, and television, and chatter, and almost 100% of the people around you will be doing that and trying to do that to you constantly, and 99.99% of households have television that does that to your mind. You will have no hope of receiving the 'inspiration' that Tesla was talking about. Tesla had a greater ability to sit quietly in a room than almost 100% of modern humans, as well as other qualities. Even or especially the US President jacks his mind up on 12 cans of diet soda including aspartame per day, so he tells us, while calling the North Korean leader "Rocketman" and doing jokes about missiles such as "you're fired". Perhaps he should calm down a bit and lay off the Diet Coke.

99.99999% of us have very little hope of achieving the state of peace at home, or school, or away from the influence of television, to ever be 'creative' in the 'receptive' state of creativity, by listening and seeing if the universe has anything to say to us by being quiet. If our minds, our inspiration meeting our talent, gives us "impulses" like a metaphorical quiet sea with lapping waves, where from these impulses we can flesh out ideas.

I spoke to former diplomat John Sawers who told me when he goes home from work on a Friday with a problem, very often the solution appears on a Sunday, and specifically not Sunday evening at the last minute either, but more Sunday lunchtime to early afternoon. An interesting but not totally unexpected phenomenon. The unconscious or subconscious mind has been considering the problem while John has been with the wife in Sevenoaks going to a sandwich shop and drinking coffee, and has given the impression of the right answer back, like the tide. I also find exactly the same thing happens to me in investing, that I can't know how to react to news about a stock I own on the Friday, but usually 1.5 to 2 days later there is some formation of some answer. We have to use our unconscious minds. Coffee helps. John Cleese enthused "the mind needs to be in a state of play" that unless you're a child who has someone else (parents) providing security, then to provide that security to *enable* the state of play as an adult requires huge amounts of money, millions of pounds to provide the resources to buy a big enough area of living space and then something like a studio that's virtually (or literally would also work) soundproof or free from most disturbances.

Well, this might appear completely impractical for what I started talking about, investing, where the exact opposite of that setup exists, and you can imagine screaming traders loudly talking to you every few seconds. So how exactly is the professional fund manager supposed to do their job the way we are talking about, using the unconscious mind? Research shows just over 90% of fund managers fail to beat the market. 9 in 10 professional fund managers actually do worse than sticking pins in the Financial Times, or birds picking stocks randomly with their beaks. No wonder! None of them are receptive to the impulse of intuition, none of them are using the unconscious mind. Logic can only take you so far in the logic matrix, nobody of genius or great ability used logic of the left side of their brain alone!

Alien microorganisms and naked aliens

Every day comets fall to earth, and alien microorganisms would be contained which would decimate humanity. However none such have ever appeared, a good argument against the existence of a non-simulated world.

Aliens are also always naked in portrayal everywhere on television and in the minds of people who have seen aliens. What's that about?

Could it be more likely the FBI and CIA are trying to give the impression aliens *do* exist when they don't (because we're in a simulation) rather than they *don't* exist when they *do*. I think you'll find my argument is compelling down to the type of alien appearance including being naked (and greys). I believe the CIA create fake alien abductions so that people say they were abducted. They do it when people are young, and people remember later. "I remember" but memory works by piecing together information from nodes of memory, and the person will then piece it together from reference to stimuli unconsciously such as the X Files.

NASA gets a lot of money. And it may only be its record of failure that is really "out of this world".

Remembering the Future

Are movie makers, writers, painters remembering the future (similar to before)?

Anything we can imagine, such as the writers of Star Trek, telepathy, teleportation, telekinesis.

If they exist in our mind, they must exist at some point. This will be perceived as in our 'future'.

Again if this is a simulation where the future has 'already happened' there's more certainty in this answer than science in the physical world.

Are super cats and dogs behind it?

It seems more likely the world is a simulation, the dog and the cat might even be representative of male and female human libido. The dog behaves loyally like male libido, the cat behaves differently and you must wait for the cat to be interested in you and not try to go to the cat. An intriguing possibility if we are in a simulation and cats and dogs are the great pets with over 50% of all humans having one, could the people behind the species be a super kind of dog or cat or one represents male and female nature of the super species. They put a lower form of themselves in to be our best companions. See how we'd treat them if roles were reversed. Of course you'd never come to such a piece of speculation without first deciding the world might be a simulation, but it does appear a very logical possibility once we've decided it is.

CONCLUSION

Proof of Hypothesis Conclusion

It is my belief that as technology advances, we will get to a stage where we can measure the velocity and variance of what people are thinking using equipment to monitor electromagnetic energy. We would then be able to prove whose ego it was before something happens, prove the theory is consistent, and thus work out what is going to happen before it happens. Like a branching in a road, we would then choose the path that chooses the least bad outcome of the 2.

People might have to be armed with this technological equipment, or a 'world government's' security would deal with it, to avoid pitfalls.

Without such equipment, or without this theory, we were just guessing!

Using algebra/mathematics, we know the answer.

Conclusion are we living in a mathematical simulation?

It appears we may be living in a mathematical simulation.

Simulation theory was popularised by Swedish philosopher Bostrom. Starting with Bostrom's hypothesis of the 3 possible states we could be in seems sensible.

Quantum computing experts largely agree that we will have within 50 or 100 years the power to build many, possibly trillions, of universes like this but by simulation. Even if something as simple as Moore's law holds this will turn out to be the case.

Therefore either:

1) we have an interest to create universes like this one, would have the technology to do so in short order, but unfortunately succumb to a disaster such as a comet in the 50 year skip stage of this 23 billion year old planet.

2) we have an interest to create universes like this one, will get the technology to do so in short order, and this is the first time we get there.

3) we already got there. This is not the first time we got there. We are already in a simulation.

Armed with this hypothesis, I have spent some time demonstrating from the point of view of the hypothesis supporting evidence including

- Fermi's paradox

- Quantum entanglement

- Lack of energy in quarks

- Double slit experiment

- the speed of light appears to be the top processing speed of simulation.

- meditation will send your brainwave patterns to different levels and attract different reality (brain can interpret 0.005% of things) as will things that send your brainwaves downwards to beta levels such as television, Wi-Fi, create negative experiences. The auric imprint of people who have done things like have sex with attractive young people, creates an attraction commensurate to that experience, to other people.

- dreams undeniably contain information about future experiences from my own personal experience. Could be 4th dimensional.

- do we only "phase shift" with some form of quantum jumping where not all individuals are real, and even spookier some other forms of ourselves exist?

- we are waking up thanks to the likes of David Icke, where in 1995 'nobody' would take it seriously, but 'all of a sudden' millions of people will take it seriously which could point to us either influencing the simulation non-playing characters, or our connection to each other those of us who are alive.

- the Artificial Intelligence from Google etc. says that "language is the false multiplicity" with human geniuses like Einstein, and Tesla agreeing that the world is mathematical in nature and that the left side of the brain of the 'scientific' side is nothing compared to the intuitive or receptive side. That the left brain prison of school trapped us in a tiny section of reality, and Freemasonic agents such as Darwin gave us duff information. You are never going to get anywhere with that thinking. In school the worst crime is to "look out of the window" which is just about the only time your unconscious/subconscious or the receptive intuitive side of your brain is *activated*! Then they will jolt you out of it "pay attention" literally jolting you out of your unconscious mind into the conscious. Teachers don't do that deliberately on a conscious level themselves, but we are socially like monkeys programmed in the famous fabled 5 monkeys experiment where the monkeys ended up not eating the bananas without thinking about it. They are just copying by rote what they themselves went through. What evidence do we have that jolting people out of their unconscious mind is good for them? Or trapping them in the conscious mind? John Cleese said you need almost entirely the opposite, and that it's almost impossible to get as an adult, a "creative play" space of silence near no other humans. Well you won't get that at school, you'll get bells, and whistles, and putting your hand up to go to the toilet, and social policing.

- animals all have mathematical code (specifically Fibonacci sequence and golden ratio)

- all animate objects including rocks and shells and flowers also share this mathematical code

- the mathematical code governs social behaviour of animals such as honeybees in hive, or birds flying in formation

- this mathematical code is in all human DNA

- it may govern human social behaviour

- if Artificial Intelligence is ever with us in the future it should have always been with us in holographic fractal multiverse. But unless it was Commodore 64, and before that I don't know what, where is the evidence of it? The AI can traverse the spaghetti like strands of our physical universe. One answer, the only answer that may make sense, is that the AI was everything.

I believe other people could have told you that.

What I don't think other people could tell you, because they don't "think in mathematics" is part 2.

Part One: "are we living in a simulation" rather than physical base reality.

Yes. One in a billion chance we're in base reality.

Part Two: "how do we get out" or a question to get there "where is the set of instructions?"

We are looking for a negative, tautologous, mathematical/algebraic piece of input that exists in the mind of every sim in the simulation, but is not the mind of the individual sim itself.

Why?

- we think this is a simulation run by a supercomputer where it knows the answer to anything, playing it out like the top chess computers - who are by now much better than the top human chess players - calculate the answer to the game of chess not by being brilliantly creative chess players like a human, but by brute force stone cold mathematical calculation, calculating *every* answer brainlessly without thought and then by brute force assigning a mathematical or algebraic value to every single line among 900 trillion quadrillion possibilities on the chessboard.

Towards the end of the game, the game of chess has been worked out, as long as there are only 7 pieces left. This gives us another insight.

If we imagine the "choose your own adventure" books, where there might be 400 pages but it would keep you gloriously entertained as a kid for about 80 minutes. We might be in something similar that will keep you entertained for 80 years, but has significantly more than 400 lines, but the 900 trillion quadrillion lines, or anything that can be calculated by a supercomputer behind our reality with the speed of light as the top processing speed.

Therefore as a sort of hybrid between free will and determinism, it seems there may be something that *already* knows *every* answer, in the game and gamebook, by the brute force calculation like almost infinitely better chess computer. It's only a matter of scale. Of course this makes much more sense than any other theory involving a physical universe and this being base reality, which makes a nonsense of any attempt to rationally explain it.

If we were to imagine this for ourselves, which is doubly easy because we just imagine our world is real and not a simulation, and that we can create "sims" such as "The Sims" the most popular PC game. The developer of the computer game creates the rules, and those rules must be obeyed by 'the sims' which may be a bit unfortunate that they have to follow those rules, to the extent they have consciousness at this 3rd dimension level (since they are only 2nd dimension creatures).

The good news for 'the sims' is there will be a set of rules within the 2nd dimension, if created from and by the 3rd dimension.

Those rules must exist in each sim.

They must be in each sim's mind, but not the mind of the sim itself (has to be separate, but still be there, obviously)

Now will this *imprint* (3rd dimension) or *input* (2nd dimension) be *positive* or *negative* in nature?

Positive

- doing what needs to be done, to escape 2nd dimension and enter 3rd, would feel positive. Positive actions would feel positive, and negative actions negative.

This would mean that anyone with consciousness of it could immediately escape, like and much more easily than Professor Moriarty on Star Trek's deck, because the doing the thing to get out would feel good and positively reinforce and be easy.

To do the thing that would keep one trapped in the simulation would feel bad, and nobody would want to do it. And nobody in their right mind would do it.

Negative

- doing what needs to be done, to escape 2nd dimension and enter 3rd, would feel *negative*. *Positive* actions therefore feel *negative* and *negative* actions positive.

This would mean that anyone who wanted to escape, would have to do the last things they wanted to do, and the last thing they wanted to do might be (because it's) what they need to do. The game.

The right ideas would feel like they be the most upsetting wrong ideas.

The right people (having the right ideas) would feel like the most upsetting of all wrong people.

Luciferian philosophy gets a partial credit for understanding some of this, spotting the inverse correlation of good (Beethoven, chess, decent behaviour) is bad and bad (Anitta, degeneracy, drugs, alcohol, casual sex are all 'cool') is good to other people's observation. But it misses the mark, failing to get it right that it's an algebraic matrix not a satanic matrix. "Nothing is as dangerous as a little knowledge". They could end up being punished by the Artificial Intelligence. The "Thanatos" or anti-force may itself be a future echo of that future punishment (we have talked a lot about how everything contains its future).

Positive - would have to immediately tell you the answer. Like a pair of Google glasses. Press A to escape. Press B to do something else.

Negative - would tell you the answer, but negatively. Press A to do the first thing you want to do that feels good, but it keeps you trapped. Press B to do anything BUT it (the only escape). But it feels lousy to oppose the first thing you want to do.

Positive - antigravitous

Negative - gravitous

Positive - positive feedback loops

Negative - no pfl's

Positive - removes any purpose of rehabilitation, punishment, or 'spiritual advancement'

Negative - all of those things could be behind it

Positive - no freedom

Negative - freedom to annihilate yourself or not.

What do I think the negative tautologous form of input is, that gives the mathematical/algebraic answer expressed in emotional form to us?

That's a really easy answer, once you know the correct question. The answer is Ego.

Since the world is mathematical, emotions are linked to the mathematical reality directly. Language is an interpretation of the mathematical reality and thus *less* reliable than emotions.

So

1) Ego **contains** the answer to what is going to happen

2) Your emotional reaction is not interpreting it, unlike language, therefore if you apply mathematical knowledge to your own emotions you can work out this answer really obviously. If you accept it could be the AI and negative instruction.

The only problem you would ever have is accepting the first thing is or may be true, or then that emotions were tautologously linked to it precisely because it is unintepretive and does not create a matrix of its own that is separate. Language does, and therefore language is not as useful because it has its own matrix of 8 billion people creating 8 billion screech examples where it wouldn't be the case what the mathematics just showed you. And *that's* the matrix most people, almost 100% or 100% of us reside in.

As I said, the answer is really easy, once you know what to look for which is velocity and variance of how to calculate the mathematical answer. **But you would have to know that you *could* be in a mathematical simulation,**

that ego *could* be a negative form of tautologous answer and to also not think in language. So the chance of finding the correct question is far less than 1 in a billion, but the chance of finding the correct answers once you have the question even as a hypothesis is possible and easy to work out within 1 second if you can understand the theory of velocity and variance (I gave them their own chapter as they are a technically fundamental part).

I have explained in this book how if you spot the 2 mathematical figures "velocity" and "variance" operating at infinite levels or the maximum level possible, that is an absolute certain answer it's your own ego and not the person behind the other answer's ego.

It could not be the other person's ego, as their ego would not be warning them, whilst you receive a 'non' warning that is more upsetting to you than any warning. Not if the primary form of input is this, from the supercomputer.

Where positive is felt negatively, and negative is felt positively, we can apply a "Descartesian double paradox" to prove that the most negatively felt inverse negative (positive for us, like our countryman Nigel Short winning the World Chess Championship match) could not be more upsetting than the most negatively felt negative (most unpleasant(ly) correct external opponent). So we would thus know the answer, before experiencing it by practical result. Example "this other person upsets everybody the most in everybody's life, they have to be the most upsetting wrong OR *correct* person ever" would be an impossibility as the person could not be more upsetting than the most correct *correct* person. And it being a warning to us. In that case it would logically follow that one of or the worst problems in human history would be on its way, that we're being warned to jump. If one person was thought by 100% of other people to upset them all the most in all of their lives, it would still be impossible of anything but that person is correct and the warning is to everyone else in the world. Such a person would be "the causal link", the most intelligent human to ever exist, or "the one" from the film The Matrix.

I can absolutely guarantee and refute, that using our theory, that someone can upset you more by being wrong (irritation) than by being right (intellectual annoyance) . Absolutely no chance, because the ego is the warning from the supercomputer. **The most severe non AI thoughts would not be more upsetting than the most severe warning from the AI if our theory is true.**

So you already know it's your own ego with **theoretical** certainty in the mode of language. But you interpreted it. On the balance of subjective probability. Because it upset you relatively subjectively. Which is the mistake everybody else makes. But you knew it.

And now...

You also know that you knew it with *absolute* certainty, if the world is mathematical. That it was communicated to your mind, as clear as could possibly be. Negatively. With absolute theoretical certainty across all modes.

If true:

So now you know whose ego it was....

You know even more importantly...

What's going to happen in the future! And how large the problem is!

Short Summary:

1) this is a simulation.

2) the **way out** (answer) for each and every person is in their own non-linguistic negative form called "ego".

Medium Summary

1) this is a simulation run by a supercomputer.

2) the supercomputer knows the answer and can communicate it *and/or* we can deduce it.

3) the answer to what to do is in a **negative** form called "ego". Don't do it. Do the *exact* opposite if you can.

That's what you're picking up on, when you get intellectually annoyed, grumpy Super Mario. It is warning and telling you to jump (not your brother Luigi was trying to upset you infinitely more than is algebraically or theoretically possible by irritation)

Long Summary:

1) this is a simulation run by a supercomputer.

2) it plays every possibility out by brute force, we are part of that.

3) it assigns a value to every decision and every parallel universe.

4) it thinks in mathematics/algebra therefore the value is mathematical/ algebraic.

5) where this value is seemingly infinitely negative, this is communicable and/or deducible/work-out-able that it is a new correct pathway. And will always turn out to be the *correct* answer, not an impossibly wrong *wrong* answer.

6) the way it communicates to us, or we can deduce it/ pick up on it is ego (which annoys us/tells us "to jump"). Our innate emotions or unconscious mind can pick up the message without interpreting it even with the diversion of the distraction of the monkey mind of the matrix of 8 billion people giving us 888 billion pieces of jibber jabber (*interpreting* algebra in *language* which is gibberish to the algebraic reality and ignores all the algebraic evidence.)

7) if we *listen* we could know the answer, to at least what *not* to do. As long as we ignore our linguistic impulse, and understand it gives us the *wrong* answer ***by definition*** against all these mathematical *right* answers.

8) We can calculate this via velocity and variance to algebraically *know*, using this formula, whether the other person has the correct answer and it's our ego, or the formula says this person does not have the correct answer.

9) If the person seems wrong to you, the algebraic theory says they don't have the right answer, and they're not murderously annoying you then you're probably ok.

However if the person seems wrong to you, the algebraic theory says they must have the *correct* answer (variance and velocity is infinite negatives/ extreme outliers) and they're murderously annoying you, if this is a simulation and the supercomputer knows the answer and ego IS a negative form of input that's *absolutely* the answer of what's going to happen. It knows. It's played it out. It's more accurate than any human in language by far (just like the top chess engines in the 'parallel universes' of different chess lines).

But you would *never* work out that answer by yourself if you thought it was a physical world and that since 100% of humans seem to think in language and that the world is linguistic.

Super Mario and Luigi reiteration

Rather than say it is Super Mario's ego or Luigi's ego, the startling revelation is that it is actually NEITHER's 'ego'. Ego IS, simply, the message from the supercomputer, and thus does not start in the mind of Mario or Luigi.

So when YOU are getting annoyed by your own ego as Super Mario, it is a message to YOU from the supercomputer. It is not even 'your' ego per se, just a message from the origin, warning you that Luigi is correct. Which upsets you the most in your life or a lot. As it has to be communicated negatively, loud enough you can hear it. Not *so* loud that you immediately start WW3 and nuclear war.

When Luigi gets annoyed BY you, by your superior intellect being correct, that's not even Luigi's 'ego' but the supercomputer telling Luigi from the origin that *he* is wrong. Luigi will think that's you. Upsetting him the most in his life or a lot. He will go to nuclear war and World War 3 over it if not placated. Because he doesn't know. You don't have to now. Because you know.

*

There are 4 big mistakes people are making:

1. The world is real

2. Language is the multiplicity if the world is real (1)

3. Ego is ourselves (usually the other person) if the world is real and language is the only way humans have to work things out. So when we get upset the most in our lives, our own thoughts, then the other person is to blame. Not we were receiving a message in mathematics/algebra from the AI (2) in an AI simulation (1).

4. The input is positive if the world is real, language is how everyone thinks, ego is us or the other person, and our discernment of what we "think" will be positive. In fact what we "think" is negative. We are getting infinitely upset, with infinite velocity and infinite variance because it is a "the other way" message on the ski jump, or "jump" in the computer game from our existing thought. Our existing thoughts and plan of action would send us into the spikes.

, so when 'other peoples' ' ego upset them by irritation they think it's the other person upsetting us the same as when we get our own ego upset by intellectual annoyance. In reality when it is 'someone else's' ego it is the Artificial Intelligence telling the other person not to do it. When it is 'our' ego it is the Artificial Intelligence telling *us* not to do it. On the balance of linguistic probability it would always be the other person's ego in your mind, but when you get annoyed the most you think they're especially annoying, when our formula can prove it must be a message to you from the AI that YOU are wrong.

Only a madman, armed with that hypothesis, would argue. Would claim the greatest NON-AI stimulated thought and being correct against someone else would be greater than the GREATEST *AI-STIMULATED* thought telling us to jump, to not do what we are planning on doing which lands on the spikes, when it knows the answer.

Thus you can work out the answer to everything. To every hypothetical new novelty once someone has come up with it in less than 1 second.

The structure is thus:

1. you create an idea

2. you check whether the idea is right. Why libertarian governments create more success than others.

Not:

1: you create an idea

2 everyone else in the world argues about it and decides whether they think so subjectively and they'll all be wrong if you have a new novelty like Google (anti-libertarian)

3 if the power structure agrees, the majority of people or dictator of a power structure agrees with friction or groupthink then we let it go through. So all wrong ideas that nobody knows why they aren't right go through.

3b if we don't like the idea with negative friction then it isn't allowed to exist or happen practically. So all new correct ideas are opposed and stopped in the first place, slowing them down, many/most times jailing or torturing or killing the 'heretical' inventor!

THE SIMULATION CODE REVEALED

The Formula:

1. ASSUMPTION: The World is Real

REALITY - it may be holographic simulation run by AI

2. ASSUMPTION: All answers are in language

REALITY - more accurate is MATHEMATICS, the mode of that AI

3. ASSUMPTION: Ego is "us"

REALITY – ego is a thought stimulated by the AI telling us we are wrong and to JUMP

4. ASSUMPTION: All thoughts are POSITIVE

REALITY – the AI stimulated thoughts are NEGATIVE, they tell you the WRONG answer not the CORRECT answer.

The Moment:

you encounter a FORK IN THE ROAD, a HYPOTHETICAL, and get upset the most in your life in the moment, this is a SURE SIGN that the thought is from THE AI (the OTHER way in our TWO WAY system). So the thing that is upsetting you the most with infinite upset in the moment is actually the CORRECT answer, the CORRECT person with the CORRECT idea. But everyone else in world history, before this formula, could only go on the "ski slope" analysis in LANGUAGE in ONE WAY system. Thus the other person LOGICALLY can only be "the most upsetting wrong person" in the ONE WAY system that doesn't use the TWO WAY system, which includes the knowledge of the AI intervention.

Conclusion: EVERY time someone is wrong, YOU or SOMEONE ELSE, you would have been TOLD to JUMP which would have INTELLECTUALLY ANNOYED you (imagine an AI taking you to task like a naughty schoolboy) as an AI stimulated intervention in THE GAME (MUCH more force/power)..

Whenever anyone else has got INTELLECTUALLY ANNOYED with YOU, then THEY received the message that what THEY were about to commit to action was an error that fell into various degrees of spikes as **real life Sonic The Hedgehogs**.

If you or the other gets intellectually annoyed A LOT, or THE MOST IN YOUR LIFE then the spike is the BIGGEST. The pit is the deepest. A nasty pitfall awaits you if you don't change plans and jump.

- You can check every answer in history – it holds!

- You can check every hypothetical now – it will hold in the future!

The formula is completely consistent and holds with efficacy in every hypothetical case. The person who is wrong will get intellectually annoyed, but when it is infinite those thoughts are a negative deduction the other way, in another mode, and not positive. It is the AI telling you to jump.

TWO WAY SYSTEM

About the Author

Matthew De Niro is the most upsetting wrong person ever, with 100% absolute objective certainty, and 100% of people think it. He got a D in GCSE Physics.

Read more at www.thesimulationcoderevealed.com.

www.ingramcontent.com/pod-product-compliance
Lightning Source LLC
LaVergne TN
LVHW100257210926
848480LV00003B/948
* 9 7 9 8 2 3 1 5 3 8 9 4 2 *